T0207218

Management for Professionals

More information about this series at http://www.springer.com/series/10101

Ferri Abolhassan

Editor

The Drivers of Digital Transformation

Why There's No Way Around the Cloud

Editor
Ferri Abolhassan
T-Systems International GmbH
Saarbrücken
Germany

Editing: Gina Duscher, Gerd Halfwassen, Annette Spiegel, Beatrice Gaczensky, Thomas van Zütphen
Translated by: Dr. Edward M. Bradburn, Stephen McLuckie, and Jessica Spengler

ISSN 2192-8096   ISSN 2192-810X   (electronic)
Management for Professionals
ISBN 978-3-319-81128-4   ISBN 978-3-319-31824-0   (eBook)
DOI 10.1007/978-3-319-31824-0

Printed on acid-free paper

This Springer imprint is published by Springer Nature
The registered company is Springer International Publishing AG Switzerland

Foreword

Apple is the most valuable company of all time. It is a technology corporation with a market value of more than 700 billion US dollars (cf. AFP, Bloomberg, dgw 2014). Apple is also the world's most valuable brand. As such, it is the most prominent example of a finding that is no longer merely a passing trend – namely, that the successful companies of today are digitalized companies. They are stirring up global markets and overtaking even 100-year-old enterprises with breathtaking speed. By the same token, every single company will be affected by this development, and every single business model will wind up being put to the test – by new and often very young competitors.

Anything that can be digitized will be digitized, and anything that can be networked will be networked. This applies to people, machines, and products alike. Software is increasingly becoming a decisive production factor, because all of these networked machines must be controlled and all the digital data must be stored, processed, and meaningfully analyzed. So the successful companies of today also have to be – or become – "software companies." And software lives in the cloud.

Digital Transformation Is an Imperative

This book aims to explore what this development means to the business leaders of today – whether they are at the head of a start-up or a long-established company. One thing is certainly clear: Digital transformation is now essential to corporate growth. What's more, Europe's future prosperity depends on successful digitalization. Until now, however, most of the biggest digital success stories have taken place in the USA and Asia. Europe in general, and Germany in particular, must not get left behind here. As a classic industrialized nation and the land of the post-World War II economic miracle, we have great opportunities available to us. But we have to seize them. Now.

Some of the foundation stones for this success have already been laid. Back in 2012, digitalization triggered a growth momentum of around 145 billion euros in Germany (cf. Bitkom 2014). A few companies were quick to take advantage of digitalization. The German mail-order company Otto is one example. Thanks to digital transformation, this long-established firm moved on from traditional catalog sales and developed into Germany's largest online retailer for lifestyle products. Otto now sells more furniture online than IKEA (cf. dpa 2015). ING-DiBa is another digital success story. Operating not even a single branch office, this financial institution – which now has more than eight million customers – made the leap to become Germany's largest direct bank by far (cf. Frühauf 2014) and the third-largest retail bank in Germany.

This goes to show that successful digital business models are out there. But that's only the beginning. The strategy consultancy Roland Berger foresees an additional cumulative potential value of 425 billion euros in Germany by the year 2025 if German industry succeeds in its digital transformation. For Europe as a whole, the researchers have forecast extra potential value of 1.25 trillion euros. But they also point out the danger of missed opportunities: If digital transformation fails, it could lead to potential losses of up to 605 billion euros in Europe (cf. Kurzlechner 2015). This can't be allowed to happen.

Many managers are already aware that now is the time to get things on track. Nearly one in two German companies is preparing itself for the fact that new technologies will eventually call its own business model into question (cf. Ernst & Young 2015). And it's true. Digitalization requires us to rethink – or at least rework – our current business models. But digitalization can also give rise to entirely new and disruptive approaches. What many successful young companies have in common is their consistent focus on customer value and the use of superior software platforms which enable them to attract a very large number of customers very quickly. This is how Airbnb became the largest hotel provider without owning a single bed of its own, and how Uber became the largest taxi company without owning a single vehicle.

Digital Transformation Is a Management Issue

The examples reveal something else as well: The pace at which companies must respond to new demands continues to grow. There was a time when the average life expectancy of a company was 75 years. Now it's just 15 (cf. Hagel III 2010). And in more and more cases, the lack of a digitalization strategy is the deal breaker. Take Kodak, for instance: The camera and photo specialist once had 140,000 employees and annual sales of around 28 billion US dollars. Kodak missed out on the digital transformation and ultimately had to file for bankruptcy. Parallel to this, a software company named Instagram became the world's largest digital photo-sharing app – with just 13 employees. When Instagram was sold to Facebook in 2012, it cost one billion dollars (cf. Thun 2014).

Digitalization is not purely an IT issue. It is strategically important and critical to business. In light of this, when it comes to the digitalization of production processes (what are known as "Industry 4.0" applications), the management or executive board is usually responsible for handling things (cf. Bitkom 2015). Digitalization has become a management issue – as it should be.

Incidentally, this applies to big businesses as well as mid-sized companies – the so-called *Mittelstand* – which are the backbone of German industry. German Chancellor Angela Merkel thinks the issue is not yet anchored deeply enough in many small to medium-sized enterprises (cf. Fietz 2015). She has also stressed that just appointing an IT officer is not enough, because the impetus has to come from the management. And she is right – after all, CEOs are the ones who must make sure their companies are agile enough to anticipate changes and react to them quickly. The CEO has to establish the parameters for digitalization so that the IT department can cope with the new requirements.

Digital Transformation Requires the Cloud

Speed and agility are key when it comes to introducing new products and processes. Cloud computing is the technological basis for this. The cloud is what makes it possible to achieve the high velocity demanded in this age of digitalization. It makes the necessary services faster, more flexible, and more secure.

The question is no longer whether the cloud should have a place in a company's strategy – the question is about the form and scope of that place. Integrated ICT providers are the obvious partner here because cloud computing requires both technological foundations and consulting expertise – "hardware" *and* "software," if you like. One essential component of this is a strong broadband network for fixed lines and mobile communication – ideally transnational, pan-European, and all-IP. This calls for IT security which is "made in Germany" and hosted in highly secure data centers. But it also calls for sophisticated IT quality management, which is vital in the digital world, coupled with the relevant experience in digital transformation.

Ultimately, every company has different requirements and different goals which must be continuously adapted to the market conditions. An ICT partner is therefore like a company's "architect of the digital future."

Digital Transformation Requires Trust

There is another obstacle to overcome, however: More than one-third of all Germans currently say that they are mostly afraid of digitalization. Only for the under-45s do the opportunities outweigh the fears (cf. Dörner et al. 2014). This is where entrepreneurs, IT service providers, and politicians come in. Through our work, we must create a sense of trust and emphasize the possibilities that digitalization offers.

A sensitive approach to data is crucial to this. Data is the raw material of the digital economy. We have to harness the masses of accumulated data and use them efficiently – but *for* people, not *against* them. Data security and data protection must therefore always be the top priority.

Thanks to its strict data protection policies, Germany enjoys a great advantage here. We should make use of it. It is good to see that Europe is finally on its way to creating a cross-border general data protection regulation. European interior and justice ministers recently agreed to reform Europe's data protection rules. This will give a considerable boost to the establishment of unified standards. This is precisely what we have always pushed for because it is an important basis for shared and secure digital platforms in Europe. In this way, we can create a real counterbalance to the strong economic regions of the USA and Asia.

This book provides insights from various perspectives into how companies can get started with their digital transformation, which factors are critical to success, and how much potential is offered by the cloud. Specific practical examples show how German and European companies can work with the right partners to shape the upcoming second phase of industrial digitalization.

The future of the German and European economy is at stake here. We are in a strong position: Germany is an industrialized nation with an outstanding reputation as a supplier to the world. The time has come to take our expertise in mechanical and plant engineering and our understanding of quality and combine them with the advantages of digitalization. We have the necessary technologies. We just need to put them to use.

The game is on – let's get the ball rolling.

Deutsche Telekom AG   Tim Höttges
Bonn, Germany   Chief Executive Officer
November 2015

References

AFP/Bloomberg/dgw (2014). *Apple ist das wertvollste Unternehmen aller Zeiten.* In: welt.de. Accessed July 27, 2015, from http://www.welt.de/finanzen/boerse/article134722868/Apple-ist-das-wertvollste-Unternehmen-aller-Zeiten.html

Bitkom (2014). *Digitalisierung schafft rund 1,5 Millionen Arbeitsplätze.* Accessed July 27, 2015, from http://www.bitkom.org/de/markt_statistik/64054_78573.aspx

Bitkom (2015). *Industrie 4.0 ist Chefsache.* Accessed July 27, 2015, from http://www.bitkom.org/de/presse/8477_82244.aspx

Dörner, S., Camrath, J., & Preuschat, A. (2014). *39 Prozent der Deutschen haben Angst vor Digitalisierung.* In: wsj.de Blogs. Accessed July 27, 2015, from http://blogs.wsj.de/wsj-tech/2014/02/18/digitalisierung-umfrage/

dpa (2015). *Otto Group setzt auf Digitalisierung – Hohe Investitionen.* In: focus.de. Accessed July 27, 2015, from http://www.focus.de/finanzen/news/handel-otto-group-setzt-auf-digitalisierung-hohe-investitionen_id_4485602.html

Ernst & Young (2015). *Digitalisierung: Wer investiert und profitiert – wer verliert?* Accessed July 27, 2015, from http://www.ey.com/DE/de/Newsroom/News-releases/20150316-EY-News-Deutsche-Unternehmen-im-Digitalisierungsdilemma

Fietz, M. (2015). *Merkel ermahnt Technologie-Feinde: Keine Angst vor Big Data.* In: focus.de. Accessed July 27, 2015, from http://www.focus.de/politik/deutschland/kongress-des-cdu-wirtschaftsrates-bundeskanzlerin-merkel-warnt-big-data-nicht-als-bedrohung-anzusehen_id_4739542.html

Frühauf, M. (2014). *Direktbanken müssen ihre Kräfte bündeln.* In: faz.net. Accessed July 27, 2015, from   http://www.faz.net/aktuell/wirtschaft/wirtschaftspolitik/finanzinstitute-direktbanken-muessen-ihre-kraefte-buendeln-13076763.html

Hagel III, J. (2010). *Running faster, falling behind: John Hagel III on how American business can catch up.* Accessed July 27, 2015, from http://knowledge.wharton.upenn.edu/article/running-faster-falling-behind-john-hagel-iii-on-how-american-business-can-catch-up/

Kurzlechner, W. (2015). *Wucht von Industrie 4.0 wird unterschätzt.* In: cio.de. Accessed July 27, 2015, from http://www.cio.de/a/print/wucht-von-industrie-4-0-wird-unterschaetzt,3107422

Thun, M. (2014). *Internetguru warnt vor Gefahren von Big Data.* In: ndr.de. Accessed July 27, 2015, from https://www.ndr.de/nachrichten/netzwelt/Internetguru-warnt-vor-Gefahren-von-Big-Data,lanier103.html

Website ING-DiBa. Accessed July 27, 2015, from https://www.ing-diba.de/ueber-uns/unternehmen/

Tim Höttges has been CEO of Deutsche Telekom AG since January 2014. From 2009 until his appointment as CEO, he was the member of the Group Board of Management responsible for Finance and Controlling. From December 2006 to 2009, Höttges was the member of the Group Board of Management responsible for the T-Home unit. In this position, he was in charge of the fixed-network and broadband business as well as integrated sales and services in Germany. From 2005 until being appointed to the Group Board of Management, Mr. Höttges headed European operations as a member of the Board of Management of T-Mobile International. From 2000 to the end of 2004, he was Managing Director Finance and Controlling before becoming Chairman of the Managing Board of T-Mobile Deutschland. Mr. Höttges studied business administration at Cologne University, after which he spent three years with a business consulting company. At the end of 1992, he moved to the VIAG Group in Munich, where he was a divisional manager from 1997 and later became a member of the extended Management Board responsible for controlling, corporate planning, and mergers and acquisitions. As a project manager, he played a central role in the merger of VIAG AG and VEBA AG to form E.ON AG.

Contents

Pursuing Digital Transformation Driven by the Cloud

Ferri Abolhassan

Could all of the estimated 85 million pet owners throughout Europe use an app to track the activities of their pets? How could smart pills improve healthcare for over 26 million chronically ill people in Germany (cf. German Foundation for the Chronically Ill 2015) by providing them with more personalized treatment? How can a firefighter quickly find the information he needs – building plans, hydrant locations, interactive location maps – at any time of day so that he can get straight to the scene and save lives? How can over 100,000 employees in a global company work together effectively across national borders and local IT barriers? The scenarios could not be more different. But they have one important thing in common: The solution relies on the cloud.

1.1   The Cloud Can Do Many Things

The cloud is the basis for the digitalized business models and processes that will play a pivotal role in businesses in the future. For we will soon be living in a world in which everything is networked to everything else. Studies (cf. Kremp 2014) have estimated that there will be over 200 billion interconnected devices by the end of the decade. Dealing with these vast numbers requires technology that is reliable and stable. The cloud can do that. The Internet of Things, Industry 4.0 – virtually all of the IT sector's recent innovations rely on businesses being able to harness the speed and scalability of the cloud. It is the backbone and the brainpower of the entire digitalization movement. It offers more data storage and data analysis capacity. It makes it possible for an almost limitless number of users to capture and analyze huge data volumes centrally.

There are some conditions, however, including fast, high-performance broadband connections and powerful, secure data centers with high levels of flexibility

F. Abolhassan (✉)
T-Systems International GmbH, Mecklenburgring 25, 66121 Saarbrücken, Germany
e-mail: Ferri.Abolhassan@t-systems.com

© Springer International Publishing Switzerland 2017
F. Abolhassan (ed.), *The Drivers of Digital Transformation*, Management for Professionals, DOI 10.1007/978-3-319-31824-0_1

and scalability, and ideally compliant with stringent national data protection and security regulations.

We can discover just how revolutionary the cloud is if we take a closer look at the examples given earlier. Because what they show is that digitalized technologies and processes are now everywhere – in retail, in medicine, in public security, in large enterprises and in industry.

- *Example 1* – "Tail" is a dog app package designed to give peace of mind to pet owners. It comprises a modern tracking and sensing device contained in the dog's collar, and a smartphone application. It lets the dog owner know their pet's whereabouts at any time and tells them if there is anything they need to do to keep their pet secure and comfortable. To ensure the availability of this kind of app, and many similar apps being used routinely every day, there is a powerful cloud solution working silently in the background. Should the need arise, it can collect and evaluate the data on millions of pets from all over Europe and supply a personalized report to each pet owner. It sounds simple, but only because the complexity is hidden from the user and handled by the cloud.

- *Example 2* – FireFighterLog is just one of many applications that have the potential to save lives. It delivers information on the fastest route to the fire, building floor plans, the locations of fire hydrants and the whereabouts of the people trapped inside the building to firefighters in real time on their wearables via the cloud, thereby giving them a 60-s time advantage over traditional emergency response systems. Another lifesaver is the intelligent pill, equipped with a chip that transmits electronic signals as soon as it comes into contact with the stomach's gastric juices. Pertinent data such as the patient's heart rate or sleeping times is transmitted to an app from a wearable sensor patch, allowing the doctor to regularly monitor their patient's state of health and, in consultation with the patient, adjust the dose of medication in real time.

- *Example 3* – The challenge of large-scale collaboration. Storing documents locally along with thousands of gigabytes of legacy data is common practice in many large enterprises. But it is not a recipe for flexibility, especially when you operate across national borders. However, thanks to the cloud, even globally active Fortune 500 companies can achieve the agility they need for success. The cloud enables fast, efficient communication between tens of thousands of employees and provides instant access to growing volumes of business data and applications.

Any list of examples showing the relevance and power of cloud technologies would be a very long one indeed. And in each example, the cloud provides the platform for digital growth, for new business models and for the kind of innovation that our economy and our society need.

As the digital transformation of business advances, it is becoming increasingly urgent for enterprises to become fully networked organizations. Entire process and supply chains between companies, partners, suppliers and customers are becoming more tightly interlinked. This is creating completely new value networks as well as products and services that are more flexible, more responsive and more carefully targeted.

1.2 Knowing Today What the Customer Will Want Tomorrow

In a world where international competition for the goodwill of consumers has never been fiercer, agility and speed are more important than ever before. Any business planning to find, win over and keep new customers had better be ready to offer them a unique customer experience. But it doesn't stop there. Companies now have to be able to predict what customers will want tomorrow and respond to their wishes even before they have expressed them. Today's big data technologies allow hyper-personalized online shopping, tailored precisely to the individual needs of the consumer – basically a combination of real-time, technology-based data analysis and interpretation and an individualized approach to customer service (cf. Zukunftsinstitut 2015).

However, what applies online must also take place offline, and customer contact has gained new significance as a result of the digitalization process. The automotive sector is typical. There once was a time when the potential car buyer, looking for a Volkswagen, BMW, Daimler or whatever, would pay several visits to their local dealer to check out the cars and get advice on what to choose. Today when they go to the car dealership, they will already have been online to browse specifications and prices and will generally be much better informed. The car salesman has just one opportunity to seal the deal: He has to offer the buyer significant added value as soon as he enters the showroom. And with the aid of IT, he can do exactly that. Omnichannel retailing and value networks offer completely new sources of information and thus new ways for dealers and customers to interact.

Using iBeacons in the showroom vehicles and a customer app, Customer Experience Management solutions (CEM) can even identify which model the potential buyer is interested in. Test drives can be booked and prices retrieved immediately and with very little effort. As soon as the vehicle has found a buyer, it is entered into an interactive service program, which makes servicing and repairs (and quotations) transparent for the customer as never before. The cloud provides the ideal conditions for building close customer relations.

1.3 "Cloudification" Knows No (Industry) Limits

Digitalization has become the new and indispensable tool for companies not only because it facilitates direct contact with the customer. Industry is experiencing a digital transformation of the entire value chain – from design to product development, manufacturing, transportation and logistics through to sales. Virtually all industries are having to adapt their business models and products as a result. And once again, the cloud is the key to success. Port logistics is a case in point. If container traffic volumes are forecast to more than double by 2030, but a major international port has already reached the limits of its capacity, the port authority will have no option but to optimize its handling processes simply to maintain its own competitive position. Digital technology can help speed up the flow of traffic and goods and shorten the handling and waiting times of container ships. It can also help port managers as well as depot, terminal and car park operators and freight forwarders to react more quickly to the prevailing transport and infrastructure situation.

Then there is the example of food logistics. While Rewe online and Bringmeister have increased their share of the traditional German retail market thanks to clever customer service initiatives like same-day delivery and multichannel models, other business models with proven success outside Germany, such as Amazon Fresh and eBay Local, are waiting in the wings ready to conquer the German market. This shows that two things are happening: The traditional players are no longer the only ones dealing directly with consumers – and logistics specialists are now coming face-to-face with the customer. If a customer orders with a cloud-based app, they can have their groceries delivered very quickly. If they cannot be at home to accept the order personally, they have the option of getting them delivered to, say, the trunk of their car. The GPS coordinates for the car are sent by phone to the supermarket delivery person and at the same time a highly secure, cloud-based authentication and authorization system gives them one-time access to the trunk.

Scalable, high availability cloud solutions allow businesses to respond to market needs more effectively than ever before. The businesses who are taking advantage of these solutions are more innovative and have the resources available to pursue their own growth strategy. In other words, IT is the business and the cloud is the enabler. But this also means that getting to grips with digitalization has become a business necessity. The next seven topics offer some guidance to those wishing to keep pace with digital developments and understand their impact upon strategy, technology and organization.

1.3.1 Everything Stands or Falls with Digital Business Models

No business process can get by without IT today. There is very good reason why six out of the ten most valuable companies, including Apple and Google, are IT companies or IT-based companies (cf. Forbes 2015). The cloud supports the formation and the business models of many new and successful companies. And the consequences are huge. Digitalization has the potential to drive economic growth and increase sales for companies all over the world. Experts at McKinsey believe that by 2025 Internet technology could boost gross domestic product by 207 billion euros in Germany alone (cf. Dürand et al. 2014). This is an increase of nearly five percent. In order to achieve this, however, the German economy will have to succeed in setting standards for digitalization and developing some powerful business models.

"Ubering" Traditional Business Concepts

Uber, Airbnb and eBay are obvious examples of how digital change has long been a motor of innovation. The digital business models of each of these newcomers put the traditional market-leading enterprises under tremendous pressure. If they are to avoid being "ubered," as management consultant Roland Berger (cf. Berger 2015) aptly described this development, the "majors" in all industries will have to be even more vigilant in the future. Because the pace is relentless. According to market and technology experts, many companies will not cope with the challenges that await them. Gartner predicts that "digital incompetence" will cause one in four companies worldwide to lose its competitive ranking by 2017 (cf. Gartner 2013). Nevertheless, this also means that the majority of businesses will successfully meet the challenge. Market participants can therefore control whether they are a winner or a loser in the digitalization game.

Staying in the game demands agility and speed in all areas, together with innovative products that harness the benefits of digitalization. The German car industry is a perfect example. The technology fitted to today's cars has turned them into mobile data centers, a concept that has not escaped the notice of others, including US manufacturer Tesla with its Model S. All vehicle functions are controlled via a central tablet. The Internet, software, sensors and applications form a central nervous system through which vehicle data travels at high speed. But this nervous system must also be extremely powerful. At the point where conventional technology solutions reach their limits, highly scalable and secure cloud solutions have become indispensable in taming the "explosion" of data volumes and providing stability, high availability, and most importantly, security in the transmission of data. It will not take until 2050 before we have mass produced cars that allow the driver to increase the power of his engine by 10 or 20 percent whenever he requires an additional burst of horsepower simply by tweaking the engine management software via the Internet. Every carmaker is expected to have the necessary "chip tuning" solution in their options catalog within five years.

1.3.2 CIO and CEO – Best Buddies?

So who will be responsible for ensuring that digitalization is prioritized within the company? The CIO and the CEO are really the ones who need to push for the transformation and development of digitalized, cloud-based business models. They will have to understand and communicate the potential of the cloud to each department. The overall framework is in place. A new generation of IT leaders is currently emerging, bringing with them a dedicated business background and a business mindset. Thanks to the talents of these CIOs, CEOs are free to concentrate on their vision for the company. Although the CEO must understand the basics of the technology, he does not have to be too concerned about the potential pitfalls because this can distract from the pursuit of the company's vision. The CIO, for his part, must support the vision but still check it for feasibility.

Is the Chief Digital Officer the New Creativity Manager?

Digital transformation is forcing companies to update IT systems they have used for many years. This is often done by highly qualified specialists without the luxury of being able to shut down the systems or relax security rules. To help them meet these challenges as they move forward with digitalization, more and more CIOs are finding it useful to recruit a CDO. Chief digital officers, many of whom have followed a different career path and often come from the digital creative industries, can strengthen the team by contributing new perspectives, ideas and solutions (cf. König 2014). The CIO has the final say, of course, because he has the ultimate responsibility of integrating new applications and processes safely and reliably into the company's IT.

1.3.3   A Two-Pronged Approach

There will always be two aspects to any transformation towards the cloud: the operational system and the new, agile IT. Despite the need for innovation, the day-to-day work of the CIO is vital to the company's effort to create value. With the exception of startups, companies are only able to jump onto the moving train of digitalization if their traditional IT operation is running securely and reliably. Nevertheless, a dual IT infrastructure gives traditional companies the opportunity to compensate for this limitation. Digital devices provide the necessary speed and the additional agility. According to Gartner, around three quarters of all IT organizations will be working with Bimodal IT (2-speed IT) by 2017 (cf. Gartner 2014) – in other words, with the traditional IT that has evolved in the enterprise over the years and the new, agile IT designed to meet the needs of the future. Instead of going to the considerable expense of rebuilding and updating their existing IT infrastructures and processes immediately, many companies find it more effective to implement innovative, parallel digital solutions (cf. Rimmler 2015), which allow them to transform their business in stages.

Daring to Look at Things Differently – The Young IT Tearaways

If this transformation is to succeed, it needs people who think digitally. It needs a new IT generation that is business focused and curious and has a completely different perspective on technologies and processes. People who look for routes well away from the established, well-trodden ones and contribute a sense of risk tolerance and pragmatism from which the company can gain a great deal. If one solution does not work, they are bold enough to alter their approach. They believe in "killing it fast" and moving on. To create maximum added value for the entire enterprise, all players are required to define the key elements of their collaboration on a regular basis.

1.3.4   You'll Never Walk Alone – Side by Side on the Way to the Cloud

Because projects have a tendency to increase in complexity, sometimes even during their initial tender phase, it is important to have a partner on the supplier side who can provide flexibility as you progress toward successful digital transformation. An external IT service provider can offer more than just a fresh perspective on business processes and IT structures; they can also offer a wide range of expertise, will already have implemented numerous excellent cloud solutions and, ideally, will be used to thinking and acting from the perspective of all stakeholders. A partner like this will offer the customer genuine added value and will recommend and implement the cloud model that best suits their individual needs.

With this in mind, a provider should be able to offer their customers a wide range of interlocking cloud services from multiple vendors. In addition, providers should have transformation experience as well as process and sector expertise in projects of all sizes. This also applies to large corporations working in sensitive industries with high security-critical requirements or for whom high availability is an important criterion. An IT service provider must be familiar with the development and operation of hybrid cloud architectures which allow business-critical processes to move between different types of cloud services as needs change.

"Simple, Secure and Affordable" Boosts Acceptability

Dynamic delivery models such as IaaS, PaaS or SaaS, embedded in the IT and business strategy and part of the quality and safety concept, are poised to become pillars of enterprise IT. But to achieve this, they must be scalable and simple, secure and affordable, and they must also meet the company's compliance requirements. Moreover, processes, roles and service levels must be discussed and tailored to the cloud, in close cooperation with the departments involved.

During these discussions, if not before, the provider must provide answers to the key issues of digitalization: Which processes and infrastructures, and which parts of the value creation process should a company transfer to the cloud? Which path is the right one? How will business-critical data be handled? And what will it cost? And then, of course, the provider will have their own questions. For example, does the company have a bottom-line or top-line strategy? Do they want to increase income and efficiency or would they rather generate growth? (cf. Investopedia 2015). The list goes on. One thing is certain, however: There is no magic bullet. Every customer is different but necessary problem-solving competence remains the same. An IT provider with a breadth of expert knowledge is well placed to supply that competence.

1.3.5   The Cloud Also Means Partnering with the Very Best

To meet the exacting demands of digitalization and stay at the cutting edge of technology and software development, technology partnerships are essential, because today's cloud computing is all about collaboration, about managing complex ecosystems that individual IT providers cannot reliably handle on their own. They therefore need to work with experienced partners. It does not matter in what country or even what continent they are based. What matters is that they collectively develop the solution that is the most beneficial for the customer.

The ideal IT provider has invested many years in building a cloud ecosystem of leading technology partners that allows them to identify and assemble the best possible solutions. By selecting the right partner, with the right expertise and a technology portfolio to match, companies not only obtain a solution that ideally fits their needs but gain the added benefit of advice in matters relating to IT innovation, transformation and data center solutions. The provider is the vital central cog here, pulling together the various strands of the project, providing quality management, catering for diverse needs and overseeing the entire value creation network. Many companies are looking for a retail solution that is straightforward, with a simple and intuitive frontend, but also a backend that meets the highest security standards.

Successful cloud solutions stand out by offering maximum cost effectiveness and easy integration into existing IT environments. A good provider with their cloud partners knows how to seamlessly integrate enterprise and IT processes. A quality management system with partner certification will be a central component of their partner program. Overall, what matters most is expertise along an open and agile collaboration environment that makes it possible to offer the best solution from the customer's perspective. Ultimately, the focal point must always be the customer and his requirements.

1.3.6 Maximum Performance Through Maximum Security

A business cloud contains a company's prize assets: its customer and production data, its strategic content and its sensitive key indicators. Organizations cannot therefore simply place their business-critical data and applications in a normal consumer cloud. They should also know which data center the provider is using to send data to the cloud. They should be safe if the provider uses a secure data center based in Germany compliant with German regulations, as the German Data Protection Act is one of the strictest informational self-determination laws. The use of personal data is strictly prohibited and requires explicit legal permission or the consent of the person concerned. In the Anglo-American legal systems, the exact opposite is true. The US intelligence services can use American terror legislation to access personal data even without a court order. The USA FREEDOM Act, which superseded the USA PATRIOT Act in July 2015, offers better protection for the personal data of US citizens in the United States, but did not change the situation for personal data held in foreign data centers. Those wishing to place their data under the protective shield of German law with the data centers of service providers based in Germany are still in good hands.

The issues of data security, data protection and compliance must be part of the "big picture" of any digitalization program: Data and applications must be kept available, information protected from unauthorized disclosure to third parties and legal considerations taken into account. This is irrespective of whether the cloud is public or private, or a mixture – an approach that is increasingly finding favor with business today. Country-specific laws and regulations insist that these standards should be an integral part of any cloud strategy (cf. Experton Group 2015). This is made possible by the use of, for example, vendor-independent cloud encryption, the keys for which are held by the user companies.

A business-compatible cloud is also equipped with sophisticated defense systems. The provider is responsible for carrying out ongoing preventive work on the security architecture to protect the infrastructure and the customer's systems. Effective IT security is constantly learning, never static and a steadfast partner from the very beginning.

1.3.7 Highest Quality as the Basis for IT Transformation and Digital Growth

The greater the influence that IT has on a company's operation and growth, the more critical the reliability, stability and agility of the IT systems are for the company's success. This means, more specifically, adopting preventive measures such as redundant technologies and taking a comprehensive, end-to-end quality management approach. The risk of outages causing billions of dollars' worth of damages, reputational risks and serious data loss make the highest level of reliability vital, and for most businesses more important as a purchase criterion than the price of the cloud solution itself (cf. PricewaterhouseCoopers 2012).

One thing is certain: IT without incident does not exist. Businesses and IT providers must therefore work closely together to manage and minimize the fallout from incidents and to ensure that the highest quality is maintained. A study by the consulting firm Sopra Steria found that one in two German companies has no contingency plan for IT security incidents (cf. Steria 2012). This is just one area where much needs to be done to improve quality.

A comprehensive quality assurance system must consider the interaction between people and technology. To maximize IT security, technology partners have to work proactively and anticipate potential problems at an early stage. Service level agreements can help define high process standards.

Contingency management should not be ignored for those occasions when IT fails to operate reliably.

A case can be made for having a "manager on duty" (MOD) on stand-by around the clock, who can be contacted in the event of a crisis and who has the authority – unlike call center agents – to put measures in place to rectify the incident immediately and professionally. And, if necessary, to involve top management to expedite decisions on budgets and resources. It is certainly true that, to keep everything running smoothly, it is essential to have clearly defined contact persons, processes and KPIs. This saves valuable time in an emergency and minimizes downtime, frustration and costs.

The objective is to maintain a highly available, secure IT infrastructure at all times and to take continuous action at technical and organizational levels to ensure maximum stability. This is a service benefit that will ensure complete and continued customer satisfaction. In a nutshell: Without quality there can be no cloud and without the cloud, enterprises have a diminished capacity to innovate.

1.4   Conclusion

Digitalization and cloud-based processes are the key business drivers of the twenty-first century. When established as a platform for business processes, they offer companies unique growth opportunities and the potential for achieving a distinct competitive position and driving innovation.

For this to happen, measures need to be put in place now. Business strategies need to be defined and forward-thinking business models developed. The roles of CEO, CIO and CDO need to be allocated. The right IT partner, with their ecosystem and cooperation partners, need to be selected. The essential factors for a successful digitalization project are quality, stability and agility. IT managers themselves should design and implement innovative solutions according to the criteria of scalability, reliability, security, affordability and simplicity. Only when these points are internalized will companies be ready for the future and ready to reap the considerable benefits that digitalization has to offer.

References

Dürand, D., Menn, A., Rees, J., & Voß, O. (2014). *McKinsey-Studie – Diese Innovationen entscheiden über Deutschlands Wohlstand.* In: Wiwo.de. Accessed July 27, 2015, from http://www.wiwo.de/technologie/forschung/mckinsey-studie-diese-innovationen-entscheiden-ueber-deutschlands-wohlstand-/9867534.html

Experton Group (2015). *Cloud vendor benchmark 2015.* Accessed July 27, 2015, from http://www.experton-group.de/research/studien/cloud-vendor-benchmark-2015/ueberblick.html

Forbes (2015). *The world's most valuable brands.* Accessed July 27, 2015, from http://www.forbes.com/powerful-brands/list/

Gartner (2013). *Press release – Gartner says digital business incompetence will cause 25 percent of businesses to lose competitive ranking by 2017.* Accessed July 27, 2015, from http://www.gartner.com/newsroom/id/2598515

Gartner (2014). *Press release – Gartner says CIOs need bimodal IT to succeed in digital business.* Accessed July 27, 2015, from http://www.gartner.com/newsroom/id/2903717

Investopedia (2015). *Definition of "Top Line".* Accessed July 27, 2015, from http://www.investopedia.com/terms/t/topline.asp

König, A. (2014). *Was macht ein Chief Digital Officer?* In: computerwoche.de. Accessed July 27, 2015, from http://www.computerwoche.de/a/was-macht-ein-chief-digital-officer,3067798

Kremp, M. (2014). *Internet der Dinge: Kühlschrank verschickte Spam-Mails.* In: spiegel.de. Accessed July 27, 2015, from http://www.spiegel.de/netzwelt/web/kuehlschrank-verschickt-spam-botnet-angriff-aus-dem-internet-der-dinge-a-944030.html

PricewaterhouseCoopers/pwc (2012). *IT-Sourcing-Studie 2012.* Accessed July 27, 2015, from http://www.pwc.at/presse/2012/pdf/studie-it-sourcing-2012.pdf

Rimmler, M. (2015). *Gartner is right. Enterprises need Bimodal IT (a.k.a 2-Speed IT)*. Accessed July 27, 2015, from http://www.kinvey.com/blog/4160/gartner-is-right-enterprises-need-bimodal-it-a-k-a-2-speed-it

Roland Berger (2015). Accessed July 27, 2015, from http://www.rolandberger.de/pressemitteilungen/digitale_transformation_in_europa.html

Sopra Steria (2012). *Pressemitteilung – IT-Ausfall: Unternehmen schlecht für den Notfall vorbereitet*. Accessed July 27, 2015, from http://www.soprasteria.de/newsroom/pressemitteilungen

Website of the German Foundation for the Chronically Ill. (2015). Accessed July 27, 2015, from http://www.dsck.de/startseite.html

Zukunftsinstitut (2015). Retail Report 2016. Authors: Janine Seitz, Theresa Schleicher, Jana Ehret; Managing editor: Thomas Huber; Published by: Zukunftsinstitut GmbH.

Ferri Abolhassan a computer science graduate, secured his first professional role as part of Siemens' R&D team in Munich. He then worked at IBM in San Jose, USA. In 1992 he joined software vendor SAP, remaining until 2001. Abolhassan held a number of senior positions during this period, including a spell as Senior Vice President of the global Retail Solutions business unit. Following a four year tenure as Co-CEO and Co-Chairman at IDS Scheer, he returned to SAP in 2005, most recently as Executive Vice President, Large Enterprise for EMEA.

In 2008, Abolhassan moved to T-Systems, where he became Head of the new unit Systems Integration and joined the T-Systems Board of Management. In late 2010, Abolhassan took on role of Head of Production, before becoming Director of Delivery in 2013. Abolhassan was appointed Director of the IT Division in 2015, overseeing approximately 30,000 employees and 6,000 customers. In addition to his current function, Abolhassan has been responsible since late 2015 for building up the new business division "Telekom Security". The new unit will consolidate the security departments from all different Group units of Deutsche Telekom.

The Role of IT as an Enabler of Digital Transformation

2

Christophe Châlons and Nicole Dufft

The digital transformation of the economy and society could be called a digital revolution – one with as far-reaching an impact as the Industrial Revolution of the nineteenth century.

This digital revolution takes a number of guises, however. It essentially began 60 years ago with the first computer. Data processing, or information technology, then gradually spread to nearly every process and every industry with the goal of automating processes and making them more efficient. This evolution was largely made possible by continuous technological development.

2.1 The Digital Revolution

One essential step was taken in the 1990s with the spread of the nascent Internet and World Wide Web. These technologies revolutionized communication both within companies and between companies and their partners, suppliers and clients. Above all, however, they changed how companies communicate with their end customers. In this respect, the development of e-commerce and e-business at the end of the 1990s laid the groundwork for today's digitalization. At the time, however, information technology was still used mostly to support existing processes, such as logistics, purchasing and sales, marketing and customer relationship management.

The digital transformation of today goes much further. Companies are now using information technologies to develop fundamentally new business models, products and services.

C. Châlons (✉)
Pierre Audoin Consultants (PAC) GmbH, Holzstraße 26, 80469 Munich, Germany
e-mail: C.Chalons@pac-online.com

N. Dufft
Pierre Audoin Consultants (PAC) GmbH, Oranienburger Straße 27, 10117 Berlin, Germany
e-mail: N.Dufft@pac-online.com

© Springer International Publishing Switzerland 2017
F. Abolhassan (ed.), *The Drivers of Digital Transformation*, Management for Professionals, DOI 10.1007/978-3-319-31824-0_2

IT innovations not only provide support, they actually enable the radical re-development of processes and value chains. This has given rise to new value networks and to changes in the structures and power relations of entire industries. The traditional boundaries between industries are becoming blurred.

Examples of this are plentiful. In the travel industry, for instance, travel agencies have redefined their added value, while transportation providers such as train companies and airlines, along with hotels and tour operators, had to adapt to the new transparency in pricing and customer satisfaction as well as to the power of portals such as Booking.com or Opodo.

Successful new companies are popping up everywhere, while existing companies have to make substantial adjustments or face being squeezed out of the market – either because they haven't changed quickly enough (if at all), or because they simply haven't managed to adapt their business models to the new competition they face from digital companies and services such as eBay, Facebook, Instagram, Wikipedia, Booking.com, Airbnb, Uber or Spotify.

Incidentally, the music industry is a good example of just how quickly this change can happen: Barely ten years ago, Apple revolutionized the industry with iTunes. But today, streaming services (such as Spotify, Deezer and the new Apple Music) are making the once-successful iTunes model irrelevant – and not just among the youth. Apple has gone from a pioneer to an imitator.

These changes are not only affecting consumer markets (B2C); digitalization is also exerting a massive influence on B2B markets. Business customers and consumers alike now expect personalized interaction via various channels, both online and offline, along with individualized, networked products and offers as well as data-based services. As a result, product manufacturers are forced to evolve into solution providers.

And the next upheavals are already on the horizon: The new technologies associated with the Internet of Things will radically change the business models in other industries in the coming years – including the automobile industry, mechanical and plant engineering, and energy.

2.2 Technological Drivers and Effects

These upheavals are being driven largely by the following technological developments:

– *Mobile technologies*, which make it possible to access information, and thus to interact or make decisions, at any time, from anywhere.

– *Social media*, which are massively changing the opportunities for interaction within companies as well as with customers, partners and the general public.

– *Analytics and big data*, which enable companies to make well-founded decisions (sometimes in real time) and to develop data-based business models.

– *Cloud computing technologies*, which guarantee highly flexible access to applications and data under reasonable financial conditions.

– And finally, the *Internet of Things (IoT)*, which promises unlimited opportunities for interaction and new business models by connecting products and sensors.

Of course, it is not the digital technologies themselves that are causing the upheavals described earlier, but rather the economic effects of how these technologies interact.

For example, the use of digital technologies dramatically increases market transparency. Never before have customers had real-time access to so much information about quality, functionality, prices, alternatives or customer service. Purchasing decisions are increasingly being made on the basis of the recommendations and experiences of other consumers. This applies to everything from books to financial products which require extensive consultation with customers. And never before has it been easier for customers to switch to an alternative provider at the click of a mouse. Customers are using a growing number of digital and non-digital channels to find information, interact with providers and eventually make a purchase. They expect a seamless, coordinated purchasing experience across all channels and devices ("omnichannel" is the keyword here) as well as equally coordinated processes for billing and logistics, for instance.

Purchasing decisions are also increasingly being made on the basis of the services that come with a product. In many cases, services have become a more important market differentiator than the product itself. This applies to networked sporting goods which use apps to analyze athletic performance, networked household goods which can be controlled via apps, and machines whose maintenance status can be analyzed remotely. Connecting applications with devices via IoT technologies, or with other users via social media, is becoming increasingly important to the added value of an offer. At the same time, products do not necessarily have to be owned anymore; instead, they can be used as a service. Sharing models now exist not only for cars and bicycles but even for complex machines.

Above all, however, customers now expect personalized experiences and offers which are tailored to their individual preferences. But because individual preferences can continually shift, companies must be able to react to customer expectations or changing demand in real time – or, ideally, in advance. At the same time, price pressure has tended to increase rather than decrease in nearly every industry. Individualized products at a low cost are only possible with a high degree of automation. In light of this, mass customization is a key challenge when it comes to differentiation in the digital age.

2.3 The Three Stages of Digital Transformation

Digitalization is fundamentally changing our economy. There are three different stages to this:

1. *The digital workplace*: The spread of smartphones and other mobile devices such as tablets; collaborative tools such as video conferencing and chat; using social networks in a corporate environment; "consumerization" or the growing penetration of technologies (hardware and software) originally designed for private users which are miles ahead of the old IT landscape in terms of their ease of use. Together, these developments have radically changed the user experience of the IT user.

2. *The digital customer experience*: The second stage no longer affects IT users in a company, but rather the customer. In a networked, digital world, where customers share their experiences with everyone and can switch to a competitor in the space of seconds, the optimal customer experience has become the key to remaining competitive. This is because a negative customer experience will directly and immediately affect a company's brand perception and sales. Companies must therefore shift their strategic focus to the comprehensive, individualized optimization of the customer experience across all digital and traditional contact points. Design plays an important role here in terms of both graphic presentation and the design of the user experience. Simplicity, intuitiveness and reactivity are key characteristics. Up until now, most companies have focused mainly on the digital transformation of their front end to the customer and neglected the integration with the back office. But an optimal customer experience demands the company-wide digitalization and integration of all processes as well as a comprehensive focus on the customer. This is because back-end processes – in logistics, accounting, warehousing or product development, for example – can have at least as much impact on the customer experience as customer-facing areas. Therefore, in addition to transforming marketing and sales, digital transformation must involve the customer-focused digitalization and integration of all front-end and back-end processes in a company.

3. *Digital business models and ecosystems*: The third stage encompasses new sales models as well as new products and new business models which often lead to new digital ecosystems. In the medium to long term, companies will not be able to withstand the growing competitive pressure on their own. This is why traditional value chains are increasingly being replaced by digital ecosystems and service networks. A variety of stakeholders from different sectors will work together in these networks to develop collaborative business models. They will share data and information (and even predictions and correlation analyses) so that they can jointly offer a better service or assert themselves against a competitor. Such digital ecosystems are already beginning to emerge, particularly in the Internet of Things – or, to be more precise, the Internet of Things and Services, because the services surrounding networked products are what offer added value for the customer. There are already numerous examples of this in areas such as the connected car for the automobile industry, predictive maintenance in mechanical and plant engineering, smart meters and the smart grid in the energy industry, and smart health via remote access to, or even the remote control of, medical devices in the health sector.

2.4   What Digital Transformation Requires from Information Technology

Digitalization opens up limitless possibilities, but it also calls for a profoundly different way of thinking and an extensive transformation of organizations, processes and corporate culture. This is because companies are facing considerable new requirements in the digital age – particularly when it comes to information technology.

2.4.1   Agility

The most important requirement – or challenge, more likely – is agility and adaptability. The pace of change has picked up dramatically in the digital world. As a result, agility has become an essential factor for success. Companies must not only identify and respond to the opportunities and risks of digitalization, they must also adapt quickly to changing market and competitive conditions. They must be able to implement, test, refine – and then potentially abandon – new ideas very rapidly. Agile IT is absolutely essential to this. The new approach is "build – measure – learn – improve", or even "try – fail – learn – improve." Contrary to the approach taken by traditional IT organizations, the goal here is not to develop the optimal system for the next ten years, but rather to implement an idea as fast as possible and optimize it in a continual learning process. Speed, reactivity and flexibility are the keywords here. And when an idea takes off, it has to be scaled up very quickly.

Even though technology can't solve every problem, cloud computing is ideal for ensuring the agility, scalability and flexibility required here.

First of all, software-as-a-service (SaaS) makes it possible to implement and launch applications very swiftly and easily. SaaS also eliminates many maintenance and operational concerns. And SaaS developments are generally based on open standards, which usually makes them easy to integrate – either with the back office or with other applications and data sources. Alongside traditional, comprehensive applications, the SaaS model is increasingly offering a number of "micro apps" which can be put together like a puzzle and expanded if necessary.

If a required application is not available as a standard in the SaaS model, the PaaS approach (platform-as-a-service) makes it possible to develop it efficiently, test it in the target environment and then – most importantly – implement and launch it quickly and smoothly. Finally, infrastructure-as-a-service (IaaS) ensures the scalability needed to carry out individual tasks such as processing and analyzing larger amounts of data.

The strengths of any cloud model include agility, scalability, flexibility, simplicity and speed – from implementation to integration, maintenance and operations. At the same time, cloud solutions must meet the requirements of security, reliability and data protection. Hybrid approaches which combine the public cloud, a hosted private cloud or even an in-house private cloud facilitate the creation of differentiated environments for different applications, users and data profiles and their corresponding requirements.

2.4.2 Ability to Innovate

The ability to innovate is another essential challenge in the age of digitalization. While many companies in the past focused on creating and marketing existing products and services more efficiently using (IT-based) process optimization, we now need a radical change in thinking: Innovations are in the spotlight, and established companies have to act like startups. A fundamental cultural shift and new forms of collaboration and leadership are required in order to mobilize all of the innovative power in a company. Organizations must revise their conventional ways of thinking and working, their traditional management approaches and their control mechanisms.

Above all, the need for rapid innovation places new demands on IT because IT is often at the heart of innovations, thus playing an important role as an enabler. The ability to innovative therefore requires the corresponding IT architectures, IT working environments and development approaches.

As a result, iterative approaches based on the lean startup concept are becoming an increasingly important alternative to conventional waterfall models. The main goal of methods such as Scrum or DevOps is to significantly boost speed and agility when it comes to the development (as well as the implementation, operation, maintenance and further development) of new products and services. At the same time, these methods demand and promote a new kind of culture which focuses on customer-centered activities, continuous change and sharing. Interactivity is the keyword – both within a company and with customers. Developers must be able to react promptly to customer feedback and the evaluation of user data, and partners and customers are frequently being included in the innovation and development process.

It is tremendously difficult to apply such methods to companies that have always been organized along conventional, industrial lines, and the only way to do so is step by step. What's needed first are incubators which test these approaches in multidisciplinary teams, develop them further and then actively support their dissemination to other areas. An IT department is the ideal organization to initiate such multidisciplinary teams and coordinate their work because IT-driven innovations are increasingly at the core of new services, products and business models in the digital age.

2.4.3 Simplicity

Another key aim of these new development approaches is to produce simple, intuitive applications and products. Simplicity and user-friendliness are demanded at every stage of digitalization, starting with the digital workplace (user experience) and going all the way to the digital transformation of the customer front-end (customer experience). No one these days wants to have to read a manual in order to be able to use a product or a website.

The demand for simplicity is not restricted to the front office, however; the back office should and must be simplified as well. Most IT departments today are struggling to maintain and operate their legacy systems, so they have limited resources available for innovation. They suffer under the complexity of their IT landscape which has evolved over the years through adaptations, expansions and integration measures. Furthermore, most systems were developed to support existing organizational units. The new IT must be lean IT, meaning that it should encompass simple, efficient, appropriate processes and forms of organization.

2.4.4   Intelligent Use of Data

Digital interaction channels are opening up entirely new ways of collecting customer information and using it to expand and optimize the customer experience. For example, the price of a product can be adapted to the competitive environment, or the right product for a campaign can be selected on the basis of the (expected) weather conditions. A logistics service provider in Germany has developed a new business model for fast-food restaurants: The provider analyzes the correlation between weather conditions, events (such as soccer games) and delivery volumes and now sells demand forecasts to its clients.

The opportunities are even greater when products themselves can provide individualized usage information via the emerging Internet of Things.

It is critical to make intelligent use of this growing volume of data, which is the basis not only for the individualized optimization of the customer experience, but above all for the optimization of processes, for making operational and strategic decisions and for business innovations. Companies must be able to centrally collect and store this mass of data from different sources, analyze it in real time and make it available for a variety of uses.

2.5   How IT Can Become an Enabler of Digital Transformation

Nearly every German company has acknowledged that digitalization is an important issue, but most companies are only carrying out isolated digital projects without pursuing a uniform, company-wide strategy. As a result, digitalization often resembles a patchwork of ambitious but uncoordinated initiatives. But companies need a digital strategy and a digital agenda in order to establish guide rails so that they don't get lost in the jungle of possibilities.

IT has to play a key role here from both a technical and an organizational perspective, because digitally successful companies are unthinkable without centrally positioned IT. Ultimately, a central authority is needed which can

- Coordinate digital initiatives on the basis of a comprehensive digital strategy and ensure that the applications developed in this context mesh with one another;

- Bundle together data streams throughout the company, orchestrate them and prepare them for further analysis;

- Maintain and monitor the observance of security, compliance and data protection regulations; and

- Coordinate and shape collaboration with technology suppliers, including hardware, software and cloud providers as well as traditional IT service providers and resellers.

But in order to fulfill its strategic role in the digital transformation and position itself as a pioneer and service provider for other departments, the IT organization itself has to change fundamentally. Much like a startup, it must view itself as an agile, interactive, learning system in which planning and control are replaced by a step-by-step approach to solving problems. At the same time, IT officers must create an environment in which this "IT 2.0" can flourish.

This calls for the creation of latitude. IT organizations that are "trapped" in operational tasks and primarily occupied with maintaining and operating an existing IT landscape and managing IT costs lack the breathing space needed to drive the necessary change. Consolidating, standardizing and modernizing the legacy IT landscape and using technologies and business models such as cloud computing, outsourcing and offshoring can create latitude. This will be financial latitude, in that the money saved on legacy operations can be applied to new projects, and it will also be personnel latitude, in that well-trained employees will be freed up from routine tasks and can work on innovative sandbox projects instead.

Large, established corporations often have a two-part system these days: a system for historical products which is optimized for stability and efficiency (system of record), and a system for new digital offerings which is optimized for innovation and speed (systems of engagement). Here, too, there is an urgent need for (at least partial) integration to prevent the creation of new silo solutions.

If they want to meet the demand for agility and innovation, IT organizations should also review their performance management and KPIs. Experience has shown that there is little use in appealing to the innovativeness, agility and cooperativeness of employees if they are managed using conventional efficiency criteria. Leadership in agile organizations is not based on strict hierarchies and micro-management, it is built on trust and latitude for the employees on the one hand, and on managers actively communicating and exemplifying the organization's goals and visions on the other. Even in an IT department, the employees need coaches, not control freaks.

Agile approaches should also include suppliers and partners. This is all the more important when it comes to establishing digital ecosystems – that is, digital value creation networks which, according to Pierre Audoin Consultants, will become very important in the medium term. Appropriately designed contracts and incentive systems are required here. Agile developments and high standards of flexibility are not fully compatible with rigid work contracts or outsourcing agreements.

Companies should check to see whether risk- and profit-sharing mechanisms can be built into their contract systems. It is also advisable for provider management to be oriented more towards the end result and the view of the end customer.

Last but not least, the IT organization must do more to embrace one particular issue: customer orientation, or a focus on the needs of the end customer. The IT organization has to position itself as the central coordinator, technical consultant and integrator of projects at the interface to the customer, and even as a trailblazer for an optimal customer experience. However, this requires that the IT organization seek out – and even moderate – dialogue with every other department (including marketing, sales and customer service). An IT organization is in the best position to develop and implement a multi-channel strategy and thus guarantee an optimal customer experience across all contact points.

2.6   Conclusion

Digital transformation is profoundly changing our economy and society. Every company needs a dedicated strategy and a digital agenda if it wants to move forward efficiently. Technologies such as cloud computing, analytics and agile methods can help implement this strategy. Speed, agility, flexibility and reactivity are absolutely essential. Above all, however, the ability to innovative is the critical factor for success in the digital age. This calls for latitude, new forms of cooperation and the integration of partners and customers. Information technologies are the heart of digitalization. Agile and innovative IT is essential to the agility and innovative capability of the entire company.

In light of this, the IT organization has to play a central role in digital transformation – and it has to change fundamentally as well!

Christophe Châlons has been the Chief Analyst of the PAC Group since 2009. After earning a degree in mechanical engineering, he began his career in 1986 as an assistant to the system manager in the R&D department of the EDF Data Center in Saint Denis, France. In 1989 he was appointed Managing Director of the newly founded PAC office in Munich. Under his leadership, PAC developed into one of the leading market analysis and consulting firms in the German IT industry. Christophe Châlons is also a member of the Management Board of the PAC Group. As Vice President Quality, he is responsible for company-wide quality assurance and customer-specific studies and consulting services as well as PAC's standard studies in the context of its renowned SITSI® program. His work revolves around software and IT services at a global level, outsourcing and STIE (scientific, technical, industrial & embedded IT), consulting projects for IT service providers and sourcing advice for IT users.

Nicole Dufft is the Independent Vice President Digital Enterprise at PAC Germany. In her more than 20 years in the IT and financial industry, she has served as Managing Director of Berlecon Research GmbH in Berlin, Senior Portfolio Manager at Metzler/Payden, LLC, in Los Angeles and as an analyst with B. Metzler seel. Sohn & Co. in Frankfurt. She holds a degree in economics and is an expert in new workplace strategies in the digital world and the digital transformation of companies. She leads consulting and research projects which focus on the optimization of the competitive strategies, market positioning and business development of IT suppliers and software providers.

The Digital Transformation of Industry – The Benefit for Germany

3

Dieter Schweer and Jan Christian Sahl

Mobile Internet, social media and digital services have become part of our daily lives. The era of the Internet of Things – the network of products and machines – is just beginning. Entire value chains are being transformed by digital technology, some of it evolutionary, some of it disruptive. Anything that can be digitized will be digitized. Are Germany and Europe at the forefront of this movement? A study by Roland Berger Strategy Consultants on behalf of the BDI concluded that the digital transformation could add around 1.25 trillion euros to Europe's industrial value creation by 2025 – but could also diminish it by 605 billion euros.

German industry boasts a wealth of innovative companies. Its outstanding competitive strength in manufacturing, logistics and science and its balance of large enterprises and medium-sized businesses makes it ideally qualified to spearhead the digital transformation movement. Many of Germany's manufacturing companies are highly automated. There are approximately 282 robots for every 10,000 industrial jobs in Germany – compared with 14 in China. Germany has numerous "hidden champions," world leaders in specialist and highly complex industrial product service systems. Moreover, Germany, along with the US and Japan, is one of the biggest manufacturers of embedded systems, those postage stamp-sized computers that manage complex control and data processing tasks in a wide range of products and equipment. Although virtually unseen by users, these systems play an important part in our daily lives. They are to be found in medical devices, washing machines, cars and production machinery. They reduce costs by around 25 percent and are responsible for about 80 percent of the product innovations in today's cars. Their share of industrial value creation will continue to rise. If the truth be known, however, comparisons with other countries, including with other EU member states, reveal that Germany has a number of weaknesses, some of which give cause for serious concern. Other regions of the world dominate so many fields of the digital

D. Schweer (✉) • J.C. Sahl
Bundesverband der Deutschen Industrie e. V., Breite Straße 29, 10178 Berlin, Germany
e-mail: D.Schweer@bdi.eu; J.Sahl@bdi.eu

© Springer International Publishing Switzerland 2017
F. Abolhassan (ed.), *The Drivers of Digital Transformation*, Management for Professionals, DOI 10.1007/978-3-319-31824-0_3

economy that technological dependency has become a real risk. For example, European companies account for only about 10 percent of the global sales revenues in information and communication technology (ICT). While the ICT market is growing at 11 percent in China and 4 percent in the US, in the EU growth is a mere 1.3 percent. The manufacturers of the best-selling IT hardware, PCs and laptops are based in the USA and Asia. The situation is no different with smartphones and consumer electronics. The biggest IT service providers and almost all of the top-selling software companies are headquartered in the USA. Only a few European companies have earned themselves a place near the top of the global rankings. And despite dominating the sector 15 years ago, today there isn't a single European company in the world's top ten cell phone manufacturers.

What has been the problem? What needs to change to make the digital single market a reality in Europe? What is working well at the moment? Do we have the right innovation culture and is the legal and regulatory framework up to the job?

3.1   Building Digital Freeways

A powerful, secure and widely available digital infrastructure is the foundation of any digital economy and society. Manufacturing, networked medical services, intelligent mobility—the success of many of tomorrow's solutions depends on fast, reliable networks. Germany should be leading the way here. Instead, in an international comparison of average connection speeds Germany is in 28th place (cf. ⊚ Fig. 3.1).

This is not solely a problem for the digital economy. Broadband networks vary depending on where you live—just like power and transport networks. And like any other infrastructure investment, investment in broadband infrastructure increases the capital stock of the economy, which in turn increases the per capita income and

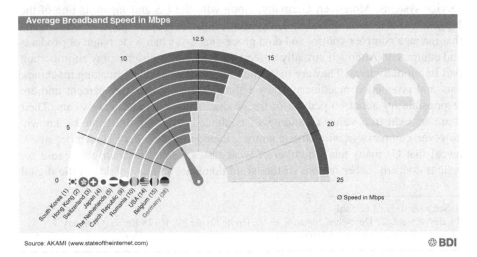

Source: AKAMI (www.stateoftheinternet.com)   ⊛ BDI

Fig. 3.1 Average Broadband Speed in Mbps

productivity within the economy. Important social challenges like the switch to renewables or intelligent traffic management cannot be mastered without a powerful digital infrastructure.

The German government's aim of achieving nationwide coverage at connection speeds of at least 50 Mbps by 2018 is a step in the right direction. In the long term, however, this can only be an intermediate step. Industry's machine-to-machine communications, the applications of the networked healthcare industry and the entertainment industry's products (video on demand, UHDTV, video telephony, computer games) require more powerful broadband connectivity. Studies have shown that, by 2025, high-volume broadband users will need an average speed of 350 Mbps. Even occasional users are forecast to need 60 Mbps. The sums required to deliver this level of service are enormous. Achieving the German government's broadband target will take an investment of between 20 and 35 billion euros. The cost of rolling out a nationwide fiber optic network is estimated at somewhere between 80 and 95 billion euros.

But the telecommunications companies will only invest in the expansion of their networks when there is a financial incentive to do so. The regulatory framework plays a vital role here. The liberalization of the telecommunications market put tremendous pressure on prices – for the benefit of many customers. Today, a national phone call costs less than three percent and some international calls less than one percent of what they used to cost. If we are to have a regulatory framework that is supportive of innovation and investment, the regulatory triangle of Germany's Telecommunications Act needs to be rebalanced. The government's three regulatory objectives include safeguarding competition, providing universal services and supporting an efficient telecommunications infrastructure. This will take more than low prices for German end consumers. It will take a powerful telecommunications industry with the ability to fund the expansion of broadband. Synergies also need to be exploited more effectively, through better coordination of construction and fiber optic cable laying projects and the shared use of other infrastructures.

Also important are carefully balanced rules on network neutrality to allow providers to differentiate their services. It must be possible for telecommunications companies to offer special services that guarantee a minimum connection speed, response time or reliability. This will provide additional resources to fund the development of the kind of digital infrastructures that are critical for the success of Industry 4.0 solutions in particular. It is clear that the digital applications of the industry of the future will not function without a guaranteed minimum connection performance.

3.2   The Web Requires Trust and Security

Unless there is confidence that data will be handled reliably and securely, many digital innovations and data-centric business models will simply not gain acceptance. Trust in digital applications requires confidence in the security and integrity of data.

Unfortunately, the revelations about widespread spying by the intelligence services have dealt a serious blow to public trust in the integrity of data (cf. ◉ Fig. 3.2). Concerns about unrestricted access to information can have a significant impact on how digital networking is accepted. The EU must work with other countries to find the answers. Personal data – as well as valuable trade secrets – must be better protected. Strong data protection and reliable data security are a competitive advantage for Europe's patchwork of different data protection regimes is economically harmful because it fragments markets, prevents economies of scale, blocks innovation, discourages many smaller companies and confuses many customers. After nearly four years of talks, the EU General Data Protection Regulation really should have been agreed by now.

Companies rely on being able to transfer data globally. Indeed, without global data transfer, networked personnel and legal departments or global R&D projects could not even function. Data-based innovations in particular need a modern data protection regime. Only by intelligently linking data (including personal data) can we deliver new, often customized, solutions to the customer – from mobility products and services, to energy utilities and healthcare services. Germany cannot dissociate itself from this trend.

A carefully balanced legal framework for data processing in Europe could simultaneously satisfy both objectives: greater trust from knowing that personal data is well protected and more innovation as a result of sensible regulations for new business models. The EU's strategy for data protection should be based on more transparency and greater customer sovereignty. In addition, the lex loci solution should apply: All suppliers to the European market, even those from outside Europe, must comply with European law regardless of the location of the corporate headquarters or the place where data has been processed. This is the only way to ensure that all EU countries will enjoy reliable standards in the future. This is an important locational advantage.

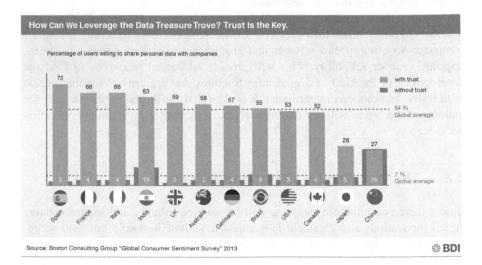

Fig. 3.2 How Can We Leverage the Data Treasure Trove? Trust Is the Key

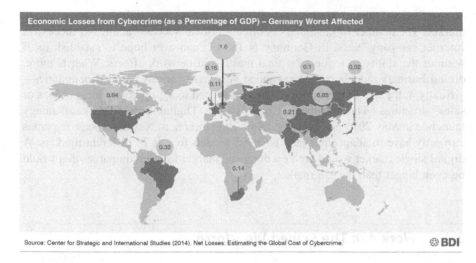

Fig. 3.3   Economic Losses from Cybercrime (as a Percentage of GDP) – Germany Worst Affected

Equally important for confidence in the digital world is the security of data and IT systems. The growing number of system interfaces crossing corporate boundaries and the widespread use of wireless communications are making attacks on corporate IT systems easier to carry out. Today, the damage done to German industry by cybercrime is thought to be around 50 billion euros annually. Approximately one in three companies have been victims of cyber attacks. Germany loses a higher percentage of its gross domestic product from cybercrime than any country in the world (cf. ☉ Fig. 3.3).

To ward off cyber attacks by organized crime and external intelligence services, government and business must work together to strengthen IT security. It is very much in the interests of German industry to protect the IT systems and networks that support their back office and production operations. Corporate security continues to improve and is subject to regular audits. However, because the issue ultimately affects all of society, close integration between industry, public authorities and academia is essential. This is why German industry has been backing the IT Security Act. Some business sectors have been subject to extensive statutory reporting and disclosure requirements at national level for some time. The German government's implementation plan (KRITIS) to protect critical infrastructures calls for efficient reporting processes both to government authorities and between companies – bilaterally as well as within the German CERT Association.

3.3   Strong Digital Domestic Market

Europe urgently needs a digital single market – a large European market in which the same rules apply, from Portugal to Finland and from Ireland to Romania. Because the Internet does not stop at national borders. Digital business models are particularly reliant on – and benefit greatly from – proximity to a large, single market. The number of customers is often a critical success factor. No successful Internet company based in Germany or the EU can ever hope to establish itself without the ability to exploit "critical mass" and network effects. What is more, digital business models can be expanded very easily across national boundaries – virtually at the touch of a button. There is no need to set up costly local offices or sales structures. The European Commission's Digital Single Market strategy launched in May 2015 is a first step in the right direction. Nevertheless, companies currently have to adapt their new business models to suit 28 different markets. A digital single market would create a domestic market for our companies that would be even bigger than the US market.

3.4   Work 4.0: The Skilled Workforce

Are Industry 4.0 factories all manned solely by robots and machines that communicate with each another and work autonomously? Unions and companies both agree that this is not what Industry 4.0 is all about. Yet there can be no doubt that digitalization is changing the world of work. For a start, the digital transformation offers workers greater freedoms, such as working hours and working models that adapt to suit the individual. Work times and locations as well as task fulfillment are becoming increasingly flexible. Secondly, the digital transformation is placing greater demands on employee skills and qualifications. There is a strong demand for well educated professionals who continually adapt their skills and qualifications to meet new and emerging technological needs. The education system must also adapt more quickly to new technologies. In order to understand the complexity in the digital industry, subject specialists and managers must be capable of thinking and acting across systems. The engineers in a smart factory need to understand not just the production processes and manufacturing technologies; they also need to have an appreciation of the underlying IT infrastructure as well as the increased security risks. Completely new, highly demanding job profiles are emerging. Because of the massive amounts of information and data being generated, we need specialists who can filter the relevant business data from the vast amount of data available. Data scientists like these are still relatively scarce. This will have to change if we are to make the most of the digital transformation's potential for growth. The fall "MINT" Report from the Cologne Institute for Economic Research revealed that German industry has a shortage of about 120,000 specialists with professional or academic degrees in the MINT specialisms of mathematics, IT, natural sciences and technology. As well as establishing long-term education and training programs, Germany needs to encourage qualified professionals to move to Germany from other countries. Any red tape standing in the way of the practical implementation of immigration law should be removed as a matter of urgency. In the future, communicating more effectively with professionals from outside Germany and managing immigration in a more targeted way will make a big difference.

3.5   Cloud Computing Guidelines

Nowadays, it is not unusual for German businesses to outsource their data processing to efficient and reliable specialist providers. In addition to managing the usual business processes, cloud services are frequently responsible for the data management of the new, data-intensive business models being used by German industry, since many of the big data analysis tools require much greater computing capacities than the average company's own IT system can provide. Cloud services are thus playing an active part in stimulating new growth in the information society (cf. ☉ Fig. 3.4).

Cloud computing also influences the added value created by industry and thus the long-term competitiveness of the German economy. The EU Commission estimates that the cloud could create as much as 250 billion euros of added value along with 2.5 million new jobs in the EU by 2020. However, before it can gain acceptance, there must be trust in the integrity and security of the cloud. Unfortunately, recent revelations about the activities of foreign security agencies have left their mark. We know from Bitkom's "Cloud Monitor 2015" that security, trust and transparency are key success factors for cloud providers. It also appears that data center operators in Germany have a clear locational advantage: More than 83 percent of the customers surveyed said that they expect their cloud provider to operate its data centers exclusively within Germany. The government needs to put in place guidelines that will create confidence and strengthen Germany's position as a favored location for data centers, which are after all the data factories of the digital economy and the physical basis of the Internet for the factories of the manufacturing sector. With DE-CIX, the world's largest Internet exchange, political stability, robust data protection (restricting access by police and security agencies) and a reliable power supply system, Germany has the ideal conditions for attracting international cloud providers. However, due to the high price of

Why Is Cloud Computing of Such Value to Enterprises?

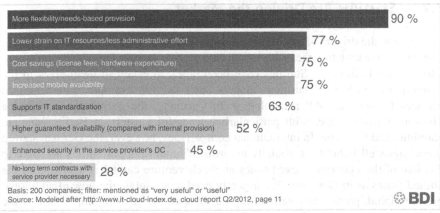

More flexibility/needs-based provision — 90 %
Lower strain on IT resources/less administrative effort — 77 %
Cost savings (license fees, hardware expenditure) — 75 %
Increased mobile availability — 75 %
Supports IT standardization — 63 %
Higher guaranteed availability (compared with internal provision) — 52 %
Enhanced security in the service provider's DC — 45 %
No-long term contracts with service provider necessary — 28 %

Basis: 200 companies; filter: mentioned as "very useful" or "useful"
Source: Modeled after http://www.it-cloud-index.de, cloud report Q2/2012, page 11

❀ BDI

Fig. 3.4 Why Is Cloud Computing of Such Value to Enterprises?

electricity in Germany, the market for data centers is actually growing much faster in other European countries. In Amsterdam, where operators pay just two thirds of the German electricity prices, data center space has increased by 75 percent since 2008. In Paris, where electricity prices are approximately half those in Germany, it was 57 percent. In Frankfurt, by contrast, the market has grown by just 20 percent since 2008.

Unlike some other energy-intensive industries, data centers in Germany are not exempt from the Renewable Energy Act surcharge. Yet power accounts for about 40 percent of the total cost of running a data center. Offsetting these disadvantages by improving energy efficiency is not really possible. According to a study by Bitkom, 45 percent of German data center operators would not rule out relocating their operations abroad. This should set the alarm bells ringing and tell the politicians that they need to do something about it.

3.6   Centralized Digital Platforms

The digital transformation of traditional industries faces a number of challenges. To successfully network industrial value chains, norms and standards need to be defined and innovative technologies, business models and new training concepts need to be implemented. In Germany, this can be done with the "Plattform Industrie 4.0" or with the "Industrial Data Space", the promising Fraunhofer-Gesellschaft initiative in which businesses work alongside politicians, associations and trade unions to try to solve important technical, legal and social problems. Similar initiatives to promote and coordinate digitalized industry have now been set in many other countries in Europe – and around the world. In the United States, for example, there is the Industrial Internet Consortium (IIC). Founded at the beginning of 2014, the organization now has more than 150 members, including a sizable contingent of German companies such as Bosch and Siemens.

3.7   Startups Are Driving the Market

Startups are the drivers of a country's innovative power. Today, we often see young companies working together with established global players. This can be a win-win situation for both sides. Startups have been quick to establish themselves in the marketplace with new business models such as, for example, Internet-based services for Industry 4.0 technologies. In Germany, the digital economy has a vigorous startup scene, with particularly successful clusters in Berlin, Munich, Hamburg and Cologne. In international comparisons, however, the German startup scene lags well behind – especially in terms of funding opportunities. Relative to the size of the economy, seven times as much venture capital is available in the United States as in Germany. We urgently need to stimulate the flow of additional private capital, particularly venture capital. Despite the growing importance of the private equity sector, the German equity market has a great deal of catching up to do to match its European competitors – not to mention the United States. This will only happen if there is a change in the statutory framework, and particularly in the tax laws. The 2008 Private Equity Act was no more than a first step in improving the funding opportunities available to young, innovative, unlisted businesses.

Dieter Schweer Master of Business Adminstration (MBA) and graduate journalist, has been a member of the Executive Board of the Federation of German Industries (BDI) since 2009, where his responsibilities include "the digital transformation of industry". Before joining the BDI, he held a variety of management positions in German industry, working for companies such as BMW AG, the Handelsblatt Group, for RWE AG as Director of Corporate Communications and for Deutsche Telekom AG as Executive Vice President. Dieter Schweer received the title PR Manager of the Year and was awarded the German Marketing Prize.

Jan Christian Sahl is a lawyer and Senior Manager in the "Digitalization and Industrial Value Chains" department of the Federation of German Industries (BDI). Sahl studied law in Bonn and Zurich. He undertook his practical legal training in Berlin, Cape Town and Malta. Prior to joining BDI, he worked for the Research Services of the German Bundestag and various international law firms and consulting companies in Berlin and Brussels. He is a visiting lecturer at the Berlin School of Economics and Law.

Theses on Digitalization

4

August-Wilhelm Scheer

Companies are alarmed by the buzzword "digitalization" and don't know specifically what to do. They are hearing about the threat of new business models, reading about the enormous sums paid for startups which haven't yet made a single dollar of profit and watching their competitors join in the chorus of scaremongers who say their sector is making the big change to Industry 4.0 or is about to be replaced by e-business. And here's the problem: All of this could, or will be, true – the massive attack on the status quo has already started! The following eleven theses on digitalization explicitly describe how a company should actually respond to the new competitive environment in order to succeed in the market.

4.1   In Praise of Slowness

Former hedge fund manager and author Nassim Nicholas Taleb uses the term "antifragility" to describe how a system can leverage an external attack in order to hit upon new ideas and take advantage of the situation. This is known as making a virtue of necessity!

But how should you actually get started if you want to come out of a situation stronger than before instead of just timidly running with the pack?

You certainly don't do it merely by copying what everyone else is doing. As a well-known philosopher once told me: "If you just hastily follow the waves of innovation, you'll come in second at best. If you want to come in first, you need the courage to go slow. Then you have to sit down with a blank piece of paper, analyze the situation, identify what's driving new technologies and use this insight to develop your own approach."

A.-W. Scheer (✉)
Scheer Group GmbH, Uni-Campus Nord, 66123 Saarbrücken, Germany
e-mail: info@scheer-group.com

© Springer International Publishing Switzerland 2017

33

F. Abolhassan (ed.), *The Drivers of Digital Transformation*, Management for Professionals, DOI 10.1007/978-3-319-31824-0_4

It can be useful to think about organizational contrasts here: Does digitalization support a more decentralized organizational structure or a more centralized one? Does it lead to more standardized or more customized products? Do customers want to take ownership of a product, or is it enough to have access rights to it?

It is also helpful (and sometimes distressingly telling) to try to answer the question of what kind of customer benefit model currently applies to one's own products or services. Often the answer is neither spontaneous nor convincing. And now this question has to be answered for digitalized products and services, too.

True innovations rarely emerge from teams because social pressure drives people to compromise and fall in line with vocal opinion leaders. So the new business model – consisting primarily of the benefit model for the customer as well as the revenue model and the resource model – must be hammered out alone, in private, at most with the aid of heated discussions and brainstorming sessions. This is the only way to create something truly new.

4.2   Carefully Sound Out the Content of Catchwords

If concepts like Industry 4.0 (I4.0) are too narrowly defined, they run the risk of being discredited because they can't prove their worth. For example, I4.0 is often applied solely to plant automation through cyber physical systems (CPS). But it should actually be used to make manufacturing more flexible, among other things. The goal is to be able to produce a lot size of one at mass-production costs. However, this flexibilization only makes sense if product development is made more flexible as well so customized versions of products can be created more quickly. This link between CPS and product development is reflected in the I4.0 concept promoted by industry associations such as the German Electrical and Electronic Manufacturers' Association (ZVEI), the German Association for Information Technology, Telecommunications and New Media (Bitkom) and the German Mechanical Engineering Association (VDMA). But even this does not go far enough. More product versions lead to more fragmented procurement processes, more suppliers and thus to more complex logistics, all of which has to be taken into account. Only by looking at the entire value chain is it possible to draw up a comprehensive model for I4.0 and a meaningful business plan for a company.

We are unaccustomed to thinking in terms of comprehensive business and value chains because this type of thinking takes us beyond the traditional boundaries between disciplines and divisions. But because digitalization makes value chains more complex, it is essential to take such a comprehensive approach. For example, the beneficiaries of a service are often not the same as those paying for the service – such as when a social network can be used for free, while revenues are generated through advertising.

4.3   Disruptive or Gradual Digitalization

When it comes to digitalization, companies can either determine which parts of an existing product or business process can be digitalized, or they can fundamentally question their existing product or business process and develop a new approach which is only possible through digitalization. The first approach leads to more gradual digitalization, while the second leads to disruptive digitalization. Both approaches have advantages which become apparent when we look at the digitalization of the taxi industry, for instance.

The mytaxi and BetterTaxi applications have digitalized (and expanded) the services offered by conventional taxi dispatchers, but they have not challenged the professional taxi business. We could refer to this as gradual digitalization. Uber, on the other hand, wants to bring down the taxi business by offering private ride-sharing opportunities through its UberPop service. Uber's business is based on the zero marginal cost principle as described by the futurist Jeremy Rifkin (cf. Rifkin 2014). This is a disruptive approach.

In a similar vein, the American company Airbnb is disrupting the traditional hotel business by providing access to private rooms, while the Internet service TRIVAGO has merely digitalized the process of finding hotel rooms.

Google is said to always take a disruptive approach. It refers to this as its "10×" vision – meaning that it wants every new project to be at least ten times better than an existing solution. When it comes to car development, Google is focusing on completely driverless operation, and it wants to flip around the current car usage ratio, where cars sit idle 95 percent of the time and are only driven 5 percent of the time. Both Uber and Google are repositioning the car. Ownership is no longer important; what matters is access to mobility as a service. Digitalization has therefore led to an entirely new – i.e., disruptive – business model.

4.4   Organizational Drivers Behind Cloud Computing

There is no such thing as an optimal organization. Sometimes a centralized organization will have advantages that outweigh its disadvantages, while in another situation the benefits of a decentralized organization will win out. Centralization has advantages in that similar activities are handled in a similar way, economies of scale can be leveraged and supervision is easier. The benefits of decentralization, by contrast, include the use of individual skills and requirements as well as more highly motivated users. Organizational tensions also exist between a strong focus on optimal processes and the utilization of resources.

When electronic data processing was first introduced, centralized systems known as mainframes were used. In the 1960s they could easily cost several million dollars. IT managers therefore aimed to make heavy use of their expensive computers. Capacity utilization rates of over 95 percent were not unusual. However, the high capacity utilization of the computer as a resource came at the expense of user processes. The time-sharing operating systems in place at the time led to long response times, and central IT departments had to manage long waiting lists of user requests for applications. User processes were therefore badly supported.

Organizational principles changed when the PC wave hit; users now had decentralized access to their own computers, they no longer had to deal with waiting times and could develop their own "applications" with office software. User processes improved considerably. But the decentralized PCs and servers of today have a low capacity utilization of sometimes just 10 percent. Process optimization now dominates resource optimization.

This low resource utilization offers potential for stronger centralization – and this is one of the drivers of cloud computing. Since the cloud can be used 24 h a day worldwide, its servers have a high capacity utilization with corresponding economies-of-scale effects. Centralization simplifies maintenance, system protection and updates, as well as training and supervision.

The goal now is to ensure that these centralization and resource advantages do not come at the expense of user processes. Cloud services must be as easy to access and system availability must be at least as high as it would be with a PC.

However, there are very good reasons to believe that it will be possible for cloud computing to combine the advantages of centralization and decentralization, as well as of resource and process optimization.

4.5   The Courage to Market Yourself

The most prominent examples of spectacular digitalization success have come from the USA (see Uber and Google). But digitalization is a global phenomenon, and every single country and company can make use of its potential and strengths to achieve extraordinary success. However, it is important for this success to be publicized internationally through PR and marketing campaigns. Germany and German industry must not acquire a reputation for having introduced the first waves of industrialization and then let digitalization pass them by. The results of research projects and excellence clusters (such as OWL = OstWestfalenLippe, Emergent Software Cluster Southwest Germany) can serve as impressive examples of digitalization from Germany, and model factories such as those operated by Wittenstein in Fellbach or by Bosch should be highlighted as trendsetting digitalization projects.

The startup scenes in Berlin and Munich are also producing some impressive results. The news that big American companies like Microsoft and Apple have bought German tech startups like six Wunderkinder GmbH and Metaio GmbH at high multiples of earnings proves that these startups are highly innovative – but the risk is that their technologies will now be refined overseas. These startups might have been able to achieve an international breakthrough on their own. For this to be possible, however, the startup scene urgently needs the support of a growth and internationalization program. The financing scene also has to be structured so that the focus is not on the exit – when investors sell the company for the best price – but rather on the sustainable business growth of the startup.

The achievements of XING, a social network for professionals, and Research gate, an Internet platform for researchers, also show that digitalization success stories are possible in the B2C sector in Germany.

In general, German companies must be more proactive in demonstrating that they feel positive about tackling the challenges of digital transformation.

4.6   Many Digitalization Obstacles Are Home-Grown

The "innovator's dilemma," a phenomenon described by Christensen (2016), is a significant impediment to digitalization in companies. It refers to the tendency of successful companies to defend their products and services against newer technologies if these technologies threaten their existing business model. Software companies that have primarily sold licenses fear that their sales will decline if they instead offer their software as a service and invoice customers based on usage. Automobile manufacturers fear that their sales will decline if they no longer sell their cars but instead bring in a car-sharing concept. Along with the fear of cannibalizing their existing business model, managers are afraid they will lose personal authority if they introduce a new business model and new technologies. If they have achieved success on the basis of their skills and experience with the existing business model, they might be squeezed out by the experts of the new era. For this reason, established companies usually hesitate to embrace disruptive digitalization and instead tend to take a cautious approach – if they take any approach at all.

Apparently serious arguments about the lack of data security offered by new technologies should also be viewed with a certain degree of skepticism. This argument is frequently trotted out when it comes to using cloud solutions, and Germany is especially sensitive in this regard.

Without dismissing these concerns, it is important to remember that if companies hesitate too long, there is a risk that both the users of digitalization and the developers of digitalized products and business models will lose ground to international competitors. This discussion must therefore be conducted with a sense of proportion and rationality. If users in Germany are too cautious, fewer offers will be developed here, and we should not be surprised if such offers are then created overseas instead and do not fit with our legal system or business regulations. Looking at this issue rationally, we can also see that the same standards are not being applied everywhere. How else can we explain why an industrial company would be skeptical of the data security of a cloud solution while simultaneously sending unencrypted CAD data between different sites by e-mail?

Government also tends to defend existing concepts rather than promote disruptive new approaches to digitalization. For instance, copyright law has prevented or impeded many new Internet services. No one takes into account that this legal construct is just a few 100 years old and arose with printing technology. But if it arose with a technology, it can be modified with a new technology. Many authors and musicians are now voluntarily foregoing copyright remuneration and offering their work for free online through YouTube and similar platforms. They have adapted their business model and are gaining more benefit from wider publicity (and the more lucrative live performances that go along with it) than they would from meager royalties. It is understandable that publishers would cling to their copyright revenues as long as possible and try to influence legislation on this. But they themselves know that they are fighting a losing battle, which is why they are building up their digital business at the same time.

Services such as Uber have also been prohibited – or at least impeded – by passenger transportation regulations. This has been done to protect the traditional taxi industry, even though this industry once unscrupulously squeezed out horse-drawn carriages.

Road traffic regulations are making it difficult to carry out extensive practical tests for driverless cars. As a result, driverless agricultural vehicles are now being used on farms and are more highly digitalized than passenger cars.

It is important to stress once again that supply and demand must be taken into account in each of these cases. If we delay or hinder the demand for digital services, there will be fewer startups in Germany and fewer established companies offering such services. Instead, these services will be developed in more receptive countries. This will result in economic losses, and it also means we will not be able to play a part in shaping these developments and will instead have to use foreign systems which might not meet our expectations.

The German government has acknowledged the importance of this development with its "Digital Agenda." The annual IT summit is a meeting point for politicians, scientists and businesspeople to discuss advances in digitalization.

4.7   Jeremy Rifkin Is Right: And Hopefully Wrong

In his bestsellers *The Third Industrial Revolution* and *The Zero Marginal Cost Society* (cf. Rifkin 2014), successful author and futurist Jeremy Rifkin highlighted some fundamental economic developments which could be brought about by digitalization.

His main theories are:

– Digitalization and general technical advances are giving rise to more and more new products which can be produced at practically zero marginal costs.

– The capitalist economic system is giving way to a society characterized by sharing and common use (the sharing economy and the collaborative commons).

– Digitalization will destroy many jobs and lead to unemployment.

It is obvious that production at low marginal costs leads to a decline in prices for products and services. In keeping with Moore's Law, it is now possible to store data and carry out calculations at practically no cost (cf. Moore 1965). The same applies to communication services (such as making telephone calls with Skype). Educational content can also be delivered over the Internet via Massive Open Online Courses (MOOCS) at almost zero marginal cost. Mobility (e.g., Uber) and overnight accommodation (e.g., Airbnb) are now available at zero marginal cost as well. This applies to renewable energy, too, and the creation of physical products by means of 3D printing with simple materials (e.g., sand).

Overall, this will lead to a sharing economy in which access to products is more important than ownership of them. In the USA, the "millennium" generation which is responsible for this new behavior is estimated to be around 80 million people. Many of them already "work" as Uber drivers or part-time consultants and use Airbnb for overnight accommodation. They have no permanent employment contract and thus no ties to a particular company; instead, they offer their services as freelancers over the Internet. This means that they produce small units of work for multiple clients at the same time. This type of work demands new kinds of social safeguards in the event of unemployment, illness, retirement or disability, an issue which is already being discussed by politicians in the USA. The flexibilization of the working world has also led to the term "gig economy." Musicians use the word "gig" to describe a performance, and the gig economy is characterized by varied, short-term assignments. In Germany, the organization and experiences of the *Künstlersozialkasse* (social welfare fund for artists) could be used as a template for social safeguards in the gig economy.

Digitalization will make many office and factory jobs obsolete. This will particularly affect medium-skilled workers. Whether this will lead to major unemployment problems is debatable, however. New digital services will also spring up, resulting in the creation of new jobs. Years of experience have shown that while technical progress can cause labor problems in the short term, it creates new employment opportunities in the medium term – as the current shortage of skilled labor shows.

It is also unlikely that the market economy will collapse. The sharing economy will certainly become a reality in some areas, but it will not dominate the entire economic system.

4.8   Technologies Are Becoming Services

The development of new digital services is an important driver of digitalization. The trend towards a sharing economy indicates that, for many products, the focus is not on ownership but on access to their functionality. This means that a product can be available on a rotating basis to many consumers. Consumers will no longer need to buy a car in order to use it, but can instead make use of the mobility services offered by car-sharing providers. In economic terms, this takes advantage of centralization and economies of scale. The Internet makes it possible to provide these services at a low cost and in a short period of time.

The tendency for technologies to be utilized as services is not new. In the nineteenth century, many craft enterprises supplied their own energy using water, wind, steam or electricity, and they even had their own water wells. Later on, energy and water supply companies emerged which offered energy and water centrally as a service and issued bills based on usage.

This development continues today and is driven by information technology. Many industrial companies are turning into service providers. This transformation has been fueled by the fact that many technical products have become increasingly difficult for occasional users to comprehend and manufacturers are better suited for operating these products. A manufacturer of agricultural equipment can offer driverless, satellite-guided harvesting services more professionally while also arranging for the crops to be transported by its own vehicles. By accessing the data from all of its devices, the manufacturer can contrast and optimize their use worldwide. The manufacturer thus has an ability to gather and analyze data which is not available to the individual user.

Analyzing large volumes of data ("big data") generally opens up the opportunity for new services and startups. While the term business intelligence (BI) was once used to refer to data analytics with an eye to past performance, direct access to current data now makes it possible to respond to events immediately or take measures proactively. Predictive maintenance is one example of this: Analyzing the current behavior of a machine – which could easily have 100 to 200 sensors attached to it for monitoring energy consumption, speed, equipment status or the attributes of the component – can reveal an imminent need for maintenance. The equipment manufacturer itself can offer maintenance services and compare the behavior of all of its products, thus gaining an information advantage over an individual user with a different maintenance policy. Startups can specialize in sophisticated analysis techniques and work together with users and system providers here.

4.9   How Do You Get Started with Digitalization?

Because digitalization affects all of a company's existing products and processes while also making it possible to develop new products, services and processes, a company-wide digitalization strategy is a complex project.

Such a project can only be realized bit by bit. The question is how these bits should be defined and in which order they should be tackled. First, the company's core processes and main products must be identified. With an I4.0 strategy, logistics, product development and manufacturing are the main processes that will be further divided into sub-processes. The maturity-level approach is helpful for prioritizing the processes to be analyzed and digitalized. To begin with, the categories of "better," "same" and "worse" are used to determine how a company's processes currently stack up against those of the competition. Then the required investment amounts are estimated. These are also roughly assigned to the categories of "low," "moderate" and "high." The competitive gap and investment categories can now be compared in a matrix. This is a good basis for discussion. It certainly makes little sense to prioritize the digitalization of a process which already has a competitive advantage and would take a great deal of investment to digitalize. Conversely, it is worth digitalizing a process that is lagging behind the competition and would require a low investment. Of course, digitalization must fundamentally improve the process in question.

A second approach (cf. Scheer 2015) compares the necessary investment with the innovative potential offered by digitalization. In this case, digitalization with high innovative potential and low investment requirements will trump digitalization with low innovative potential and high investment costs.

These financial considerations still leave room for entrepreneurial and visionary ideas which are hard to express in numbers. This makes digitalization exciting, and it is what creates winners and losers as well as opportunities for startups.

4.10 The World Is Becoming Flat

One common way of reducing the complexity of a problem is to structure it. For example, when you want to write a book, you first draw up a hierarchical outline and base your writing on that.

Simplifying hierarchical structural models exist for companies, too. For industrial enterprises, these models may take the form of hierarchical levels: the field, where the sensors are attached to a machine component; the machine controller, such as a PLC; the station, such as a machine; the workshop, such as a flexible manufacturing system; and the company. The information architecture and flow of information are also geared towards these hierarchical levels. For instance, production orders are generated by the ERP system on the company level, transmitted to the workshop and sent from there via manufacturing execution systems (MES) to the production facilities. These assign the appropriate NC programs and activate sensors and actuators. The return flow of information is also traditionally based on this hierarchy.

The influence of the I4.0 concept makes it difficult to maintain this hierarchy. If a sensor determines that a machine is no longer working with the necessary precision, a message can be sent directly from the machine's PLC to the order management system in the ERP system indicating that the customer order will be delayed. At the same time, it can request a new order which requires less precision. Horizontally, the controller can ask the PLC of another system if it can take over the order that was started – that is, whether it can guarantee the required level of precision. In principle, every component can communicate with every other component, making hierarchies largely obsolete.

The same applies in the fields of logistics and administration. Here, too, the flow of information is geared towards processes, not hierarchies. Employees from different departments can communicate with each other directly, without directing the flow of information through their respective superiors.

With the mytaxi app, passengers communicate directly with the taxi driver, without involving a taxi dispatcher.

A distributor's vehicles can independently report their expected arrival time at the customer's front gate, without involving the distributor's dispatch department or the customer's incoming goods department.

In terms of systems technology, users employ all kinds of devices, including smartphones, tablets, notebooks and computers. The application systems must be adapted to this mixture of channels and always be responsive. Because they must continually react to events, the systems should effectively respond in realtime and according to defined rules. In this respect, too, they must always be responsive. The business-process-as-a-service (BPaaS) architecture developed by Scheer GmbH follows these principles (cf. Scheer 2015).

Information and communication structures are increasingly taking the form of horizontal networks in which every node can communicate with every other node through its logical link to the business process. As a result, the business process – and not the hierarchical structure – becomes the guiding principle. Standards such as the OPC Unified Architecture for securely exchanging data between different systems and hierarchies are helpful and necessary here.

In general, the world is becoming "flat" and more complex. Because of this, it is increasingly important to document (model) business processes on the type level and to track and store them on the instance level. All business processes – with their data, statuses and changes throughout their life cycle – can be stored in a process memory so that they are available for analysis and compliance with governance criteria. Transparency is becoming the key element of control.

4.11 Software Is Eating the World

This comment by Marc Andreessen, co-founder of Netscape and developer of the Mosaic browser, hits the nail on the head. Software is the most important resource in the digital world. Software is revolutionizing existing processes and bringing about new services. It is also being used to manage and analyze data. Control over software is thus becoming the most important corporate resource. Industrial companies that have always thought their construction and production engineers were their most important knowledge resource are starting to change. They are hiring more and more software engineers, purchasing software companies or spinning off their own IT departments into independent companies in order to give them a clearer focus and expose them to the innovative pressures of the market.

Other industries are moving in the same direction. Publishers are expanding their Internet business and becoming software companies. And faced with the existential threat of Internet commerce, retailers are having to take action themselves in order to avoid being squeezed out.

Digitalization will be the biggest revolution of the twenty-first century.

References

Christensen, C. M. (2016). *The innovator's dilemma: When new technologies cause great firms to fail*. Boston, MA: Harvard Business Review Press.

Feld, T., & Kallenborn, M. (2015): Prozesse zeitnah umsetzen – In 4 Schritten zur Business Process App, http://scheer-management.com/whitepaper-prozesse-zeitnah-umsetzen_in-4-schritten-zur-business-process-app/

Moore, G. E. (1965). Cramming more components onto integrated circuits. *Electronics, 38*(8), 114–117.

Rifkin, J. (2014). *The zero marginal cost society*. New York: Palgrave Macmillan.

Scheer, A.-W. (2015). Industrie 4.0: Von der Vision zur Implementierung, http://scheer-management.com/whitepaper-industrie-4-0-von-der-vision-zur-implementierung/

August-Wilhelm Scheer is one of the most influential scientists and entrepreneurs in the field of business information and the software industry in Germany. His books are considered standard works on business process management, and the ARIS management method he developed for processes and IT is used by nearly all DAX companies and many SMEs, even internationally. He is the founder of a successful software and consulting company which he actively supports. The companies in the Scheer Group include Scheer GmbH, imc AG, e2e Technologies, IS Predict and Backes SRT. As an entrepreneur and key player in the "Industrie 4.0" and "Smart Service World" projects of the German federal government, he is helping to shape the digital economy.

The Cloud in the Driver's Seat

5

Guido Reinking

No other industry is changing as quickly as the auto industry. And nobody – apart from the IT and telecoms sector – is pushing harder toward the cloud. Automobile manufacturers and marketers are under great pressure to become mobility service providers. This change is being driven by trends such as the electrification of the drive train as well as the political and social pressures to make mobility as sustainable as possible. And for some time now, innovation in the industry has not come from the core areas of mechanical engineering, thermodynamics, chemistry or physics, but from electronics and software development. The German auto makers are aware of this. But are they doing anything about it?

Not without some justification, the industry now expects advances of historic proportions – at least in their manufacturing processes. It is no coincidence, therefore, that even the CEOs of European automakers were showing an interest in robots at the recent IZB International Suppliers Fair. More specifically, they were looking for robots capable of working alongside people. After all, robotics technology is part of a phenomenon that the Massachusetts Institute of Technology (MIT) calls "The Second Machine Age" and in Germany is known as Industry 4.0. But even this, the biggest development in the automotive industry since the introduction of the assembly line and lean production, is only part of an even bigger change. As a result of the digitalization of all areas of our lives, the mobile Internet and the widespread networking of technical devices throughout the value chain, Germany's key industry, responsible for one in seven of the country's jobs, is facing a crucial test. New suppliers from the IT industry with new business models are forcing their way into the automakers' established customer relationships. Some have even started building cars, although Elon Musk, the founder of electric car pioneers Tesla, considers these to be more like an "iPhone on wheels" than a means of transportation.

G. Reinking (✉)
Guido Reinking Automotive Press GmbH, Alte Bahnhofstraße 23b, 82343 Pöcking, Germany
e-mail: gr@guidoreinking.com

© Springer International Publishing Switzerland 2017   45
F. Abolhassan (ed.), *The Drivers of Digital Transformation*, Management for Professionals, DOI 10.1007/978-3-319-31824-0_5

These "mobile devices" must be designed, produced, programmed, marketed and serviced differently from the cars we are familiar with today. This is not just because the IT industry is keen to increase its share of the automotive industry's value chain but because a new generation of customers is coming through, and they have different requirements. In the future, simply selling cars will no longer work as a business model. Customers want mobility and are prepared to spend money to get it. But investing 20,000, 40,000 or even 60,000 euros in something that stands unused outside your door for 22 h a day seems pointless, particularly to the younger generation of motorists. This is one of the catalysts of recent success stories being written by mobility providers like Car-to-go, DriveNow and Uber.

No wonder that the impact of the digital world on the automotive business has become a burning issue for the industry. Today, every mid-size car carries more software on board than an Airbus A 320. This has one serious consequence: Although all the automotive manufacturers have gone shopping in Silicon Valley and are acquiring software and hardware companies and vast numbers of programmers, they must appreciate that, without the help of external IT companies, they will not get very far. These specialists are needed for tasks such as integrating new apps, or tracking down large numbers of developers from all over the world and quickly checking their competence for the automotive sector. "We can never keep up with tens of thousands of programmers," said Wolfgang Ziebart, who was the Technical Director at Jaguar Land Rover until March 2015 and a former CEO of Infineon. "The automakers have to successfully win over the application developer community to their cause. This is fundamental to success."

5.1 The Pressure Is Enormous

The "big boys" of the Internet are already poised to overtake the established automakers if they fail to become more agile in the IT stakes. Google has already launched its "self-driving" car. It conforms to young people's notion of buying mobility as and when you need it rather than owning your own expensive vehicle. Apple's plans for a car have attracted the attention of Wolfsburg, Munich and Stuttgart, although VW, BMW and Daimler as well as truck manufacturers like MAN have been road testing autonomous vehicles for some time now. According to a survey carried out by vehicle finance company Lease Trend, one out of every four of today's car buyers can imagine purchasing or using a Google or Apple car. What makes the prognosis even more painful is that these customers are mostly male, above-average earners – the traditional target demographic of the German premium car brands.

But perhaps the threat from Silicon Valley is this: If you don't let us into your cars with our business models, we'll just build our own. One thing is certain: OEMs around the world – from Ford to VW to Toyota – are working feverishly on their own IT strategies, apps and products. According to a study by Roland Berger Strategy Consultants, the global market for driver-assisted system software alone will be worth as much as 20 billion US dollars by 2030. Market research company

Juniper Research predicts that the number of apps used in vehicles worldwide will grow to around 270 million by 2018. In 2013 there were only 54 million. And according to another survey, car buyers are willing to spend more than 3,000 euros for safety-enhancing advanced driver assistance systems. The question is, who will develop these software solutions and capture the lion's share of the business – the car makers and their suppliers or the big Internet and technology companies? Tesla manager Philipp Schröder is in no doubt: "The companies in Silicon Valley rightly believe that they will decide the future of the automobile." That is why Tesla sees itself more as a software developer than as a car manufacturer.

How have the manufacturers been reacting to this? Daimler CEO Dieter Zetsche recently recalled how a new technology had turned the rules of the mobility market on their head once before. It took only 25 years for steam engines to be replaced by diesel and electrically powered locomotives. None of the steam engine manufacturers survived the transformation of engine technologies in the rail industry. "That will not happen to us," promises Zetsche. His company has a clear "commitment to innovation."

BMW was also quick to recognize the need for change when it set up the Project-i with its own in-house team to develop concepts for the new world of the automobile. Cars like the i3 and the i8 are the most visible manifestation of a new era, although the newly developed IT architecture concealed beneath the carbon-fiber bodies is likely to be of greater significance for the company's future than the electric cars themselves.

5.2   Industry 4.0: The New Machine Age

Human-machine interaction is an important aspect of the networked, highly automated production environment known as Industry 4.0. The sheer pace of digitalization and networking is leading to an innovation leap in robotics technology. Sensors, actuators, controllers and image processing are becoming more powerful and less expensive. "Robots that can help production staff by taking over heavy physical work will be a common feature of the factory of the future. Their strengths of power and mechanical precision perfectly enhance the human characteristics of flexibility, intelligence and sensitivity," says Harald Krüger, CEO of BMW since May 2015.

The fact that more and more robots are taking over jobs that were previously carried out by people is not just due to the employers' concern for the health of their employees. The German babyboomers, born during the post-World War II baby boom, will be retiring over the next ten years. When they do, BMW, Daimler and VW will find it difficult to find qualified people to fill the posts they leave vacant. "Over the next few years, the children of the Economic Miracle will be turning 60 and then going into retirement," says Horst Neumann, Volkswagen Board Member with responsibility for Human Resources. "Between 2015 and 2030 we will see an extraordinary number of employees leaving the company due to retirement." This is why VW is looking into the possibility of replacing people with robots without increasing the unemployment rate, which is what happened in many companies in the 1980s and 1990s.

Also, automation brings another benefit. Labor costs in the German auto industry, at about 40 euros per hour, are three to four times higher than in Eastern Europe and China. A robot can do an assembly job for between 3 and 6 euros an hour – including maintenance and wear and tear – and its deployment will ultimately secure the jobs of a fair number of human workers. VW manager Neumann believes that the use of robots increases the demand for skilled workers, foremen and engineers with IT skills. In the future, staff will not be required to carry out heavy physical work, but brainwork. Volkswagen has had robots working side by side with people for some time now in its various plants.

The German automakers are confident that they are leading the world with Industry 4.0 – particularly since they also have the support of some highly capable machine tool and IT companies. SAP and Telekom have already formed a consortium to set standards for the new automation and networking processes in manufacturing. The idea is to develop a closed process chain, from development, to the component supply companies and manufacturers, right through to service. This will provide the first end-to-end documentation of when, where and how a part from a particular supplier was installed in a particular vehicle, thus simplifying the work of the dealership's service department. According to a PwC study, the digitalization of the automotive industry's value chain will increase both horizontally and vertically over the next five years from the current 20 percent to over 80 percent. Volkswagen has already introduced a product data management software throughout the Group that depicts a continuous digital process chain from the design of a component through to its manufacture.

But modern production processes like these are not without their risks. Industrial espionage is a more serious problem for the automotive industry than most. Intelligence about what new technologies and models the industry giants currently have under development could be worth a lot of money to some of their competitors. Developers are therefore skeptical about sharing their data with OEM suppliers and the production department. This skepticism is what led Daimler to become one of the first German companies to employ its own team of experienced hackers. Working from a secret location, they carry out attacks on their own employer in order to quickly identify vulnerabilities in the company's internal network. "Although most hacker attacks last only a few hours, it often takes months before they are detected". Daimler plans to become much more adept at this game of catch-up and is also imposing more stringent security standards on its suppliers. One of the issues it is looking into is the security of its web-enabled machinery. Robots are considered particularly vulnerable to attack because they are not as well protected as office workstations. Security experts are already concerned that hackers could use this vulnerability to disrupt the production of entire plants, causing damage worth millions of euros. It is every production manager's nightmare.

5.3   Networking: Cars Driving in the Cloud

Networked assembly shops, development centers and OEM suppliers are not the only places that need protection from hacker attacks in Industry 4.0. In the future, every car that is permanently connected to the Internet – whether or not it is traveling on the road – will be a target for computer criminals, viruses and trojans. BMW has learned this to its cost, and even the successful Tesla S electric car from the USA has already been cracked by hackers. They were able to get into the car using the keyless entry system and the mobile app, open the doors and operate the horn and the lights. This is a frightening scenario, particularly for the self-driving cars that the manufacturers and component suppliers have been working so hard on.

Since Bosch introduced the first electronic fuel injection system in the 1970s, cars have slowly been evolving into traveling computers. Today, the average luxury car is equipped with 70 control units connected to three km of cable weighing a total of 60 kg. Until a few years ago, the IT was only connected inside the car and not to the outside world. This has now changed radically.

80 percent of new cars are already networked. In a few years, it will be 100 percent. Networking is integral to fleet management solutions like Car Energy Manager, which connects fleet operators but also individual vehicles with the charging infrastructure available en route to optimize their control. Another reason networking has become so important is that vehicles are increasingly becoming places for "connected life and work." Even today, customers take their digital identities with them into the car via their smartphones. For the automakers, who have always differentiated their products on the basis of their mechanical characteristics, this marks the dawning of a new era. Rather than horsepower, driving comfort or wide tires, what the young target group wants is to be able to use at least some of their cherished smartphone apps in the car.

Automobile manufacturers have to respond to this, or they run the risk of being left behind. "We spend three years developing a new car, five years producing it, and it then spends ten years driving on the roads," says Wolfgang Ziebart, former Engineering Director at Jaguar Land Rover. This is the auto industry's dilemma: No matter how intelligent the infotainment solution, a car will never be up-to-date when a dozen new generations of smartphones will be released during its lifetime. The solution that this long-established British manufacturer found for this problem is now shared by nearly all automobile manufacturers. Instead of selling the customer an expensive infotainment installation with GPS navigation, a large screen and a digital radio, which can easily account for 20 percent of the price of a small or compact car, they provide a connection from the driver's smartphone to a screen in the center console. This enables the customer to use all the non-distracting apps that have been approved by the manufacturer. This is why Jaguar only installs in their cars "those functions that are needed when there is no smartphone on board."

Drivers can use the smartphone's Incontrol apps to navigate, listen to music, manage their diary, book a hotel room or retrieve messages. Thanks to its link to the car, the smartphone always knows the car's location, whether it is locked up, or how far the car can travel on the current tank of fuel.

Volkswagen, Mercedes, BMW and Audi are also pursuing a strategy of smartphone integration. The industry has recognized that networking the automobile requires new business models. The strategy of selling outrageously expensive infotainment solutions, which are less capable than any iPhone or Android phone and which quickly become obsolete anyway, is no longer sustainable. The ability of easily updatable systems like Mirror Link (smartphone connectivity), CarPlay (Apple) and Android Auto (Google) to effectively integrate the smartphone in the car's infotainment system offers the auto industry a magic bullet.

The update capability of IT systems is a key concern for manufacturers. Because new car buyers are not the only ones who want a car that is up-to-date. New cars will suffer a huge loss in resale value if the car's IT solutions look as ancient as an old cassette deck next to an mp3 player. Just a two percent drop in the resale value of the trade's used car inventory equates to a loss for the industry of around one billion euros (Source: own calculation). Even now, navigation systems whose maps have to be updated with complex and expensive DVDs are barely able to command any premium at all. On the contrary: An obsolete GPS reduces the price of a used car. The car of the future will update its software via the Internet. Online navigation allows permanent access to the latest maps, including realtime traffic information.

The automakers' efforts to establish their own business models in order to stay in control of developments recently prompted a fierce takeover struggle for Nokia's Here mapping service. Lasting several months, it involved a consortium of German premium manufacturers – Audi, BMW and Daimler – and non-auto industry bidders including Facebook and the taxi service Uber. Here is considered to be the only service provider capable of seriously competing with Google's mapping service. This was a battle for data. And Here has a lot of data. Each road is linked to 400 items of information – from the number of lanes, to speed limits, through to restaurants and shops. The information is stored in Nokia's Location Cloud data center ready for retrieval. It is clearly not possible to handle such large amounts of data in the car, let alone keep them updated. The car of the future will therefore run not just on the roads but also in dedicated car clouds.

Unlike current car navigation systems and smartphone GPS apps, which only display a small portion of the available data, the networked, highly automated autonomous driving cloud platforms collect, process and aggregate a variety of sensor data and environmental information and can make this information available in realtime to large numbers of vehicles. It is therefore no surprise that "big data", the constantly expanding volume of data generated by vehicles on the move, is keeping the industry on its toes: Volkswagen has set up its own data lab in Munich specifically to deal with big data. Big data offers the possibility of many of the business models that marketing professionals could once only dream about: Motorists being guided to restaurants which pay for the privilege, for example or guidance to the nearest available car park. "A car can detect where the free parking spaces are in a town and pass this information on to other vehicles," says Ulrich Eichhorn, Technical Director of the VDA. Thirty percent of the traffic in cities is cars cruising for parking, says Eichhorn. A system that shows drivers where the nearest free parking spaces are could prevent much of this traffic. "A large automobile manufacturer could provide that information exclusively to its customers." Or sell it to third parties for a fee.

Business models like these are currently firing imaginations in the industry. Some are based on advanced driver assistance systems which – like AutoApp launched at the IAA – also provide location-specific driving recommendations in realtime along with a variety of other services such as weather, service station and emergency services information. There is another reason why the automobile manufacturers have no intention of letting Silicon Valley gain ownership of the screen in the car without a fight. "It is the manufacturers who decide which operating system is used in their cars and there is no reason to believe that this will change in the future," says Audi Chairman Rupert Stadler. Given that many of the in-car features affect safety, maintenance and driver assistance, the industry will not be told what to do. This is particularly true for the most complex of all the electronic assistants, the driverless car. The launch of Google's self-driving car caused the established automakers in Europe, Asia and the United States to redouble their work in this area.

No one in the industry is prepared to give a definite date for the arrival of fully piloted driving – even though the technology will soon be ready. The first hurdle is the lack of clarity with regard to liability. Who pays when a driverless car causes an accident – the user or the manufacturer? Some technological hurdles have already been overcome, however. The autopilot cannot be allowed to make mistakes when driving the vehicle. The rule is therefore: Quality of development has priority over speed of development.

Driverless cars must have a reliable, fast Internet connection if they are to travel safely in environments other than near parking garages or downtown areas. They also need "artificial intelligence." A driver knows intuitively that there is always a risk of a child leaving a group of children and running out onto the road. Because of their experience, people have a huge advantage over machines, which may be able to react 20 times faster but remain blind in many situations despite their laser scanners and image analyzers. The Centre of Automotive Research at Stanford University has worked out that it takes seven years from gaining a license before a driver can anticipate traffic situations safely. Yet people cause 90 percent of all traffic accidents. Nobody would allow that kind of an error rate in a robotic car.

5.4 New Technology: New Skepticism?

The insurance companies are keeping a close eye on these developments, even though they continue to base their new business models on the number one cause of accidents – the driver. Some insurers are now starting to offer "pay as you drive" premiums. A black box in the car constantly monitors the driver's actions and the vehicle location. Anyone who is prepared to abide by the rules or who as a newly qualified driver avoids high-risk activities such as Saturday night cruising pays a cheaper premium. Since the provider in question was given the "Big Brother Award" for its obsession with collecting data, however, the insurance industry in Germany has grown more cautious and is now reluctant to offer insurance policies of this type.

While the Americans and Chinese are quick to see the opportunities these new technologies offer, the Germans are more concerned with their risks. More specifically, the still unresolved matter of data privacy, which puts restrictions on many business models. The data that is already being generated by every networked car is certainly being eyed with great interest by many companies.

But who owns the data? Even the lawyers cannot agree. While consumer organizations claim that "motorists basically own the data in their cars," Kassel-based law professor Alexander Roßnagel writes in a paper: "Automotive data is intangible information and therefore not subject to the rules of ownership and possession."

When workshops service cars nowadays, they read out the vehicle data and error codes. This data is already being passed on online. This is why Mercedes development chief Thomas Weber promises, "When developing the fully networked vehicle, we were mindful of the issue of data protection from the outset. The car of the future will include an increasing number of IT-based features. It will therefore have to offer more than just safe and reliable transportation: It will also have to guarantee the security of its data." Careful and secure data handling is considered to be a key factor in the acceptance of new technologies.

As the industry leaders are well aware, the more benefits data processing can offer the motorist, the greater its acceptability will be. Surveys have revealed that a huge number of motorists believe in the importance of passing on data that contributes to traffic safety. They have no problem with data that alerts other road users to hazards, congestion and accidents. This could easily change if the driver starts to see unwanted offers in his display from restaurants, shops and service stations that are aware that he is in the vicinity.

On the whole, however, car owners are willing to allow their vehicles to be networked if it leads to an improvement in comfort and quality of life. A case in point is the "Smart Home Integration" feature on the BMW Connected Drive platform, which automatically increases the room temperature in the driver's home as soon he approaches. Similarly, who could argue against a vehicle that can detect wear and tear and arrange with the workshop to have the brake pads exchanged? Such services are already being offered as independent aftermarket solutions.

A new era is dawning for the more than 33,000 automobile workshops in Germany. Experts from the Volkswagen Data Lab in Munich have used big data technology to develop a model for forecasting spare parts requirements. To do this, they looked at more than 32 million records and developed an algorithm that is able to predict the parts required for an entire range up to 15 years after production has been discontinued. IT can now even predict when a vehicle needs to be serviced.

Vehicle data can also be used to achieve a long-held dream of all automobile vendors: producing an accurate needs analysis. What vehicle and what engine does the customer actually need to meet his mobility needs? And would he be better off with a gasoline, diesel or electric vehicle?

5.5   Even Car Dealerships Need to Reequip

But before those plans for the future are implemented, these car dealerships will have to get to grips with the digital marketing and sales channels that are already with us today. And in order to maintain their customer relationships, the dealerships will also have to increase their use of IT with solutions such as Product Lifecycle Management (PLM), Customer Experience Management (CEM), and so on. Cloud services covering the entire lifecycle of customer relationships, available around the clock via a mobile platform, will connect the customer with the automobile manufacturer, the auto dealership, the vehicle of his choice, and the dealer workshop.

Large volumes of data that can only be processed in the cloud and that require fast, stable data connections are rapidly changing the auto trade landscape. In their expensive, prestige city locations, Audi, BMW and Mercedes now show most of their new models on large flat screens and tablets – in addition to a small number of actual cars. All the vehicle data for these models has been digitally edited. Audi City in London has set aside 17 terabytes on their servers for the purpose. "The auto trade," says BMW Board Member Ian Robertson, "is changing as rapidly as the products it is selling." BMW calls this process "Future Retail" – the approach to displaying products virtually that has already become the norm in many of the showrooms of the Group's brands.

Digital sales is not the only area where you find huge volumes of data: "The quantity of data that a self-driving car produces can quickly reach the multi-terabyte range. This is too much for the vehicle to process comfortably and certainly too much for it to store," says Thomas Schiller, Partner in the Deloitte consulting firm. There is still no solution for the infrastructure investment that will be needed, says Schiller. Storing the data in the car would currently fill the entire trunk. Despite this, in-car computers are becoming increasingly powerful: Processors with nearly 200 cores, as announced by Audi at the CES in Las Vegas, were previously used only in supercomputers.

A number of ambitious projects have drawn attention to one area where networking falls short: the infrastructure. Take the European launch of eCall, for example. This system, which automatically transfers an emergency call on the EU-wide number 112 to the nearest control center, was supposed to have been launched in 2015 but has now been postponed until 2018. Unfortunately, many EU countries simply lack the technology needed to receive the emergency signal, which contains the vehicle's GPS information, the number of occupants and the direction of travel. The signal is triggered as soon as a vehicle's airbags deploy.

Even faster mobile networks are needed to ensure that networked mobility has sufficient bandwidth and capacity. At the moment, LTE is the fastest standard but in 2020 we will see the arrival of the fifth generation (5G) offering data rates up to one GB per second – ten times faster than the current LTE standard.

5.6   Conclusion: An Industry in Transition

German automakers and their suppliers are making good progress with Industry 4.0. However, the IT industry has the edge when it comes to networking the car and delivering applications to the car from the cloud. It remains uncertain, therefore, whether the mobility products of the future will come from Stuttgart, Bavaria or Wolfsburg – or from Silicon Valley. In the major and large cities, where half of humanity currently lives, vehicle ownership is in decline. In those places where there are high sales volumes and driving involves more stress than pleasure, competitive advantage is no longer measured in horsepower but in gigabytes and megabits per second. The Internet and the cloud have changed the way we obtain our information, in the same way as the car changed our mobility. The result will be automobiles that drive themselves, communicate with our immediate environment and with other vehicles via fast links to the cloud and are emission- and accident-free.

References

"autogramm" guest contribution by Horst Neumann. Accessed October 6, 2015, from http://autogramm.volkswagen.de/11_14/aktuell/aktuell_04.html

Auto Bild Marktbarometer (2015). Accessed October 6, 2015 from http://www.axelspringer.de/presse/AUTO-BILD-Marktbarometer-2015-Connected-Car-Ausstattung-fuer-Autokaeufer-immer-wichtiger_22815663.html

Juniper Research (2014, May 27). Connected cars – Telematics at a crossroads. Accessed October 6, 2015, from http://www.juniperresearch.com/whitepaper/connected-cars

FAZ Forum. Accessed October 6, 2015, from http://www.fazforum.com/industriegesellschaft2015/150304_FAF_Huether_IW.pdf

Frankfurter Allgemeine Zeitung, print edition, 04/06/2015; online. Accessed October 6, 2015, from http://blogs.faz.net/adhoc/2015/04/06/apple-auto-interessiert-reiche-junge-maenner-1003/

KPMG Cloud Monitor (2015). Accessed October 6, 2015, from https://www.bitkom.org/Publikationen/2015/Studien/Cloud-Monitor-2015/Cloud_Monitor_2015_KPMG_Bitkom_Research.pdf

Manager Magazin (2014, December 16). Car wars. Accessed October 6, 2015, from https://heft.manager-magazin.de/digital/#MM/2014/11/129945030

Roland Berger (2015). "Studie Automatisierte Fahrzeuge" 2nd Quarter 2015. Accessed October 6, 2015, from http://www.rolandberger.de/media/pdf/Roland_Berger_Index_Automatisierte_Fahrzeuge_Q2_2015_d_20150820.pdf

Speech by Harald Krüger, Member of the Board of Management of BMW AG, Production, on the occasion of the production of the three millionth MINI at the Oxford plant.

Wirtschaftswoche (2014, July 26). Accessed October 6, 2015, from http://www.wiwo.de/
    unternehmen/auto/cyberabwehr-bei-daimler-interne-hacker-spezialeinheit-attackiert-firmeneige
    nes-netz/10250764.html
Zeitschrift für Straßenverkehrsrecht. Issue 08/2014, p. 16.

Guido Reinking is Editor-in-Chief of the New World Mobility trade fair platform, Managing Director of the Guido Reinking Automotive Press media agency, and works as a freelance writer for publications such as the business magazine "Capital." As a journalist, he has spent two decades observing the automobile industry, initially for the "Welt am Sonntag," then for the "Financial Times Germany," and after that as Editor-in-Chief of the trade journal "Automotive Week" from 2006 to 2014. A particular focus of his reporting work has been the globalization and digitalization of the automobile sector. Reinking is an acknowledged expert on the automobile industry and regularly comments on industry trends on the ntv news channel and on RTL.

The Cloud in Practice

6

Frank Strecker and Jörn Kellermann

The cloud is a trendsetter, an enabler of new business models and the engine driving digitalization. But what does a cloud model capable of actually leveraging this new potential for businesses and the economy at large really look like in practice? An awareness of digital strategies in the minds of decision-makers is just one side of the coin – although it is a very important one. Beyond this, however, the right organizational setup – both technical and strategic – is also essential, and this is an issue that affects both IT service providers and all other companies alike. Partnerships (both inside and outside the company's own sector) are a key element here, as is having the right sales strategy across new digital channels. Last but not least, a foundation must also be laid that enables future-proof digital models in the first place: data center technology in the form of virtual IT resources, provided as infrastructure-as-a-service (IaaS) to appropriately serve the principles of simplicity, security and affordability. Here, the factors of availability and scalability also play another important role. Digital sales concepts, partnering, IT delivery models and security and/or data protection issues are therefore topics that must be scrutinized closely in the context of new ideas and innovations before these can grow into successful business models. This chapter takes a two-pronged look at these topics.

F. Strecker (✉)
T-Systems International GmbH, Fasanenweg 5, 70771 Leinfelden-Echterdingen, Germany
e-mail: Frank.Strecker@t-systems.com

J. Kellermann
T-Systems International GmbH, Heinrich-Hertz-Strasse 1, 64295 Darmstadt, Germany
e-mail: Joern.Kellermann@t-systems.com

© Springer International Publishing Switzerland 2017

57

F. Abolhassan (ed.), *The Drivers of Digital Transformation*, Management for Professionals, DOI 10.1007/978-3-319-31824-0_6

6.1   Joining Forces, Leveraging Strengths: Partnering in the Cloud

The comprehensive phase of transformation that first forced telecoms and IT companies to adapt has long since reached other types of businesses. Sooner or later, the digital transformation will come to affect every company in every industry. For IT companies, however, it's the chance of a lifetime: IT is no commodity, but is the foundation of new and successful business models. And this is where cloud is critical: It is the technological basis for the agility that today's markets ask for – and which tomorrow's markets will demand. Whether we are talking about interconnecting industrial machinery, managing large volumes of data or setting up global platforms for thousands of employees worldwide: Companies' handling of the digitalization challenge depends heavily on the availability of scalable, highly flexible systems. Whatever their industry, only the cloud can pave the way for digitalized business processes, products and corporate strategies that "think digital" – and thus ultimately uncover new potential for sales. While it certainly creates a platform that makes it easier to reach, understand and network with customers, the cloud also presents companies with a compelling challenge: They must modernize and stabilize their IT so it can guarantee both speed and scalability when providing these innovative services. Not least because target markets need to be addressed via these digital channels with tailor-made offers for innovative products. This cannot be done in isolation but needs the right ecosystem and thus the right partners. Forced to go it alone, even an IT company wouldn't succeed in this role for long – as a standalone provider, so to speak. Partnering is therefore a key element in the sales strategy of companies involved in digital market developments.

6.1.1   Focus on IT and the Customer

As we have already seen in the preceding chapters, IT (and thus cloud) is a crucial element of today's business processes and models – and even more so for those yet to come. Online banking, driverless cars, efficient production – even basic business success is now dependent on IT. As are our customers themselves. Cloud-based solutions are now an indispensable tool for sales channels and customer touch points. And the modern customer is now also very active online. In this context, digitalization has long been working to shift the balance of power. For many modern value chains, it is now the customer that calls the shots. While our customers are "always on" and very well informed, the business world often still lags behind – or isn't as up-to-date in the areas where it counts. There is too little known about target markets and the types of products they are using. This is a knowledge gap that has to be bridged. To do so, companies can take a leaf out of the customer's book and make the most of digitalization by using the cloud itself as a sales channel and talking to the target market in its new hangout (so to speak). And by bringing the right specialists on board when they do so. A cross-sector approach must also be taken: Even long-standing, well-established companies need to offer target markets extra incentives on new channels and woo them with innovative products. Accordingly, digitalization doesn't merely require a transformation of one's own business models but also the sales channels, customer communication and partnership models. So what does this mean in practice for a company's sales strategy?

The first step in this process is to establish a solid understanding of current developments in the market. Not least because digitalization both at home and in the workplace has brought changes that materially affect company orientation and demand extraordinary flexibility from every modern business. This can be summarized as three key trends that currently affect the market:

- Increased transparency: Price comparisons, product/service details, customer opinions, etc. can all be found online. Customers can research all of the relevant features for a product down to the last detail and then simply pick out the best offer from those available. The concept of *the* supplier or product is history.

- Standards compliance: Fast-moving markets and new concepts in communication – such as company-wide collaboration systems or the global rollout of virtual workplace solutions – mean compliance with standards is significantly more relevant (for reasons of cost-effectiveness alone) than was the case even just a few years ago.

- Product diversity: Competitors are increasingly offering their products globally via new channels like online marketplaces. Companies must be much more flexible and agile when responding to the dynamic markets this creates.

As these trends indicate, increasing demands for market agility and impact require a highly diversified sales process and strong partnerships. More than ever before, teamwork is the best strategy for handling the requirements of modern customers.

6.1.2 Catching Up With the Customer – With Omni-Channel

The days of the traveling salesman with trench coat, hat and oversized suitcase are long gone. As are those of traditional sales roles, too – whether in field sales or as a retail salesperson – at least in this particular form and as the sole customer contact. In the future, the sales process will dispense with this "one of a kind" model: Instead, the strategy will be to respond directly to customer needs, to be as "always on" as customers themselves. Take consumer electronics, for example: A couple of decades ago, a TV, laptop or hi-fi systems manufacturer might have received customer feedback about the product – whether praise or criticism – from a salesperson in an electronics dealership. Weeks or months later, or even on his or her next store visit – assuming the customer stayed loyal to the brand. And the trade press also published product reviews. But the tempo of customer feedback has now accelerated massively and is more business-critical than ever before. The wider scope of these discussions now encompasses Internet forums, Facebook or personal blogs and is backed by photos, videos and approving "likes". While this can be to a manufacturer's advantage – especially if experiences are positive – it is also a standing, 24/7 invitation to criticize or propose specific product improvements. These channels not only give consumers access to a brand's public image but also put retouching tools into their hands. The company concerned has to respond, and with the same agility on the exact same channels. And not merely by posting a comment on a blog but also, if need be, by personally adapting production to the new demands or developing entirely new products – which are then, in turn, presented across all available channels. As a next step, this also means that companies need to see customer feedback in a new light, however. No longer merely influencing the strategies developed within After Sales or Customer Relationship Management (CRM) but passed on to production, upstream suppliers or internal Research & Development (naturally while ensuring strict compliance with data protection regulations).

In this customer communication context, an omni-channel presence is the order of the day. An online shop or salesroom alone is simply no longer sufficient. The 2016 Retail Report neatly summarizes the problem by redefining "PoS" as "Point of Situation". When retail (for example) needs to get the message across to its customers, it is no longer just the "how" but the "where" that is relevant. This requires (for example) sales spaces that are more personal, more flexible and more adaptable (cf. Zukunftsinstitut 2015). What is more, a package of perfectly coordinated activities spanning each and every channel – as innovative and intelligent as possible – is required. The cloud forms the bridge to these new channels. It makes apps available, offers a platform for anonymous trend analysis based on social media comments, keeps online shops online even when visitor numbers surge and provides a foundation for the telepresence or web conferencing solutions that are now a key customer relationship tool in the context of B2B. In a very real sense, the cloud *is* the proximity that is necessary for customer acquisition and loyalty: It not only facilitates a seamless shopping experience for customers but simultaneously enables a wider understanding of customer needs. Companies are already well aware of the need to take appropriate action here: In 2015, the average sum CEOs at the 250 largest retailers ploughed back into the optimization of omnichannel fulfillment was an impressive 29 percent of available funds (cf. PWC 2015). And that's certainly just the beginning.

This is what the new customer experience providing complete peace of mind could look like: After configuring his dream car on the PC at home, the customer uses his tablet to add details en route, checking these with the manufacturer directly via chat. On the journey home, he uses the public screen at the bus stop to go through options for interior styling, reviewing mock-ups of colors and materials.[1]

---

[1] A digital campaign run by SEAT at bus stops in Barcelona and Madrid: After installing the mobile app onto their device, customers can then style the interior of the SEAT Ibiza in realtime on a 75-in. screen at the bus stop.

A few days later, he visits the car dealer's showroom to take his car for a virtual test drive. With the last few details sorted out, the salesman can actually offer him a number of extra features that he hasn't even thought of: The dealer now knows his customer better than ever – and in ways that simply weren't possible before. And that's not all. After the car is sold, a new kind of customer relationship management steps in: The cloud now networks the vehicle with the dealer and the repair shop, so the next service is scheduled entirely automatically and predictively – i.e., before parts start to show signs of wear. In fact, the vehicle's probably a self-driving car – thanks again to the cloud.

The first "digital showrooms" from companies featuring customer interaction show that this is a concept that works. The Audi City showroom in London has rewarded Audi with a 70 percent boost to sales figures (cf. Binder und Mortsiefer 2014). BMW is also increasingly investing in showrooms that turn the car driver's visit into a unique experience – and 44 percent of customers have already stated they would be happy to purchase their next car online (cf. Gabler 2015). Mercedes has also made its mark here, with its "Mercedes me" concept, which includes both an online shop and flagship stores in major urban centers – opened in 2014 in Hamburg and 2015 in Milan. Stores are also planned in Moscow and Tokyo, for example (cf. Behrens 2015).

But the consumer sector isn't alone in requiring rapid, accurate solutions as responses to customer requirements: The principle also applies in B2B, where an omni-channel presence is once again the tool of choice. The range of options offered by an online sales portal for B2B solutions, for example, lets users configure and manage their products while providing a direct channel to an expert. The manufacturer, in turn, can use the portal to create targeted offers or send a sales expert to a customer on a demand-only basis (rather than posting field sales staff up and down the country). The goal is the same as before: using first-class care and cross-channel support to grow alongside one's customers.

6.1.3   New Models of Collaboration: Team-Based Customer Acquisition

In the process of reaching and impressing customers – and ideally securing their long-term loyalty – omni-channel integration can only take you so far, however. Outstanding agility in the market also requires the pooling of resources. In the "old world," customers were used to choosing between just a handful of options. Market power was shared between monopolists and well-established companies. Today, however, the global economy means products from around the world are only a click away. And cross-border business is a self-evident strategy. The erstwhile office phone – communication channel #1 for decades – has long since been supplanted by sophisticated, cloud-based infrastructure such as telepresence systems and high-availability lines, carrying unimaginable quantities of data. While companies source hardware, software and services from multiple manufacturers, IT providers offer a single-source model. That's the cloud concept.

And this concept highlights a very important fact: Just because an IT company can provide the fastest network doesn't necessarily mean that it develops state-of-the-art hardware and software or knows the right answer to specific conditions that apply in other countries. Others might be better equipped to do so. Yet together, everyone can be the best. Partnerships are therefore the be-all and end-all in the world of digital because a truly integrated solution that offers the customer real value needs teamwork and a pooling of resources. IT service providers also need to allow for partner models in their corporate strategies – and with IT manufacturers and newly formed startups alike. Only then will they be able to offer their customers sound, reliable and novel solutions. This starts with the software and hardware in the data center and extends to custom development work with non-sector customer businesses. In the Connected Car model, for example, IT providers themselves become parts suppliers – and are likely to be working alongside other competitors. Odd? Not at all: Since digital models require us to jettison traditional ideas about the competition.

The upshot is that boundaries between companies and specific sectors are now increasingly blurred. The cloud concept is gaining ground in each and every industry, and partnering must become an integral part of every future corporate strategy – practiced not merely at IT service providers, but with them. Companies are picking up on this point: In one survey, over 80 percent of companies asked agreed that systematic implementation of the digital transformation is possible only in tandem with professional partners (cf. Crisp Research 2015). Ultimately, the goal here is to satisfy the most demanding technological requirements while complying with stringent data protection regulations – while simultaneously developing and further extending core competencies. This goes further than software developers partnering with hardware manufacturers to offer users a better IT solution: Beverage producers are also teaming up with vending machine makers, who in turn bring IT providers on board, to ensure that customers can always be served the right drink in the right location. The goal is to identify personal preferences as soon as possible – the Holy Grail is to do so before the preference is even stated. Activities in this field are driving entirely new B2B2C models. Coca-Cola offers a recent example: With Freestyle, consumers can mix their perfect drink literally on demand by using one of the 24,000 interactive beverage dispensers now available on the US market (cf. T-Systems MMS 2015) – with an iOS/Android app and interfaces to social networks naturally all part of the package. This effectively brings the drinks factory to the customer.

Yet these partnerships promise long-term success only if companies back the right horse in their collaborative ventures. The market is highly dynamic, and the strengths of its many participants are continuously engaged in a survival of the fittest: Those who pick the right partners will simply outlive the rest. Accordingly, companies need to surround themselves with the right ecosystem. A look over the consumer's shoulder reveals the importance of being part of the relevant platform: Facebook, messenger apps, Instagram. These platforms offer very little room for other models to attract user interest. Things are no different in the world of business: Only the right kinds of partnerships generate new ideas and products for the ever-demanding users – and thus supply a coveted entrance ticket to market developments on a global scale. Anyone who is not fast enough here and doesn't have the right ideas at the right place and time can quickly drag down other partners and even bring the entire industry sector to a standstill. Only one other company on another continent needs to be quicker off the mark and they are then free to corner the entire market.

6.1.4 Conclusion

The digital transformation is blurring the lines – between industries, between internal company units, in customer communication and in the sales strategy. This is creating new challenges and offering many new chances for companies and the economy at large. To avoid losing customers despite the agility of contemporary markets and instead respond optimally to their wishes, companies must promote well-conceived omni-channel marketing and the search for a well-structured partner ecosystem to the very top of their agendas. The cloud and the IT service provider are fundamental links in the chain here: They lay the groundwork for tomorrow's customer relationships, for as-yet unheard-of technological innovation and the future development of entire industries.

6.2   Cloud Operations: What Are the Real Priorities in Practice?

As Apple has shown us, digital growth is always powered by IT – and this is virtually synonymous with the cloud. With its characteristic attributes of high availability, flexibility and scalability, the cloud has shifted technological progress into overdrive – and yet we are still nowhere near its theoretical top speed. One thing is certain: We know that the cloud's particular properties create the environment needed for networking things together, for ensuring that networked things can communicate with another, and for keeping networked products and equipment available – which in turn optimizes business processes to generate new business models. Cloud is thus the engine that drives digital transformation. And as with any engine, it needs the right chassis to actually put its horsepower to work. Companies now need to ask themselves how cloud technology can be integrated into their IT strategy to turn this new potential into profit. Which cloud infrastructures and which applications really create added value?

6.2.1   A Mere Formality

The first step is taking stock: "Is my company 'cloud ready'?" And: "Which of my applications, infrastructures and processes should be moved to the cloud – and where is this less advisable at first? Is it better to manage my own cloud or have it managed by others? If management is going to be handled in-house, do I have the right numbers of properly-trained personnel to do so?" And: "What kind of costs are we talking about here?"

Once these questions have answers, the next problem is the selection and definition of cloud models: public, private or hybrid cloud? "Which of these is suitable, and which data is best held in which type of cloud?" "What would be the optimum 'cloud mix' for my company?" At first glance, the public cloud seems to have a lot going for it: greater innovation, more agility. But how can business-critical applications and sensitive data be protected? After all, the private cloud offers both electronic and physical layers. Or: Maybe hybrid? This alternative combines both types of provisioning to exploit their advantages, using the public cloud for less sensitive data or processes and reserving the private cloud for business-critical operations. This option is increasingly popular: In 2015, Technology Business Research forecast annual growth of 50 percent for the corporate hybrid cloud (cf. Gaudin 2014). Yet hybrid cloud computing also means greater complexity, of course. So how can this multi-cloud mix be managed efficiently? Above all, it's important for companies to work out a tailor-made strategy for the migration to cloud infrastructure. The only cloud concept that promises long-term success is one that has been carefully tailored to corporate goals and internal requirements. Choosing the right partner to sound out what really is the best cloud strategy is therefore a crucial decision. Not least because an off-the-shelf cloud is as unlikely as the name suggests.

6.2.2   The Cloud Needs a Centralized Platform

The difficulty of full-scale in-house management, even for large corporations, is well illustrated by the example of the hybrid cloud, where complexity is king and scalability is everything. While working with a service provider brings obvious benefits, what qualities should the perfect partner bring to the table? Experience, first and foremost: Only a partner who has guided enough cloud projects of all shapes and sizes to a successful outcome can manage the mix of public and private with the efficiency needed, and has the relevant know-how for transitioning the customer to the cloud in time, on budget and at the right quality. And only the early adopters of virtualization are now able to offer the innovative, powerful data center and network infrastructure equipped to handle the challenges of the future.

The foundation should therefore be standardized, dynamic cloud platforms that offer variable performance profiles to ensure rapid access from anywhere on the planet, and regardless of whether the services used are hosted by a private, public or hybrid cloud. Platforms should also be flexible and easily scalable to ensure they can adapt to any size of digital transformation. The advantage of such standardized, dynamic platforms is that their cloud hosting ability isn't limited by the constraints of time and place but follows the principle of "any service on any server at any time." This means uniform access to all cloud-enabled offerings, such as platform-as-a-service (PaaS), software-as-a-service (SaaS) and infrastructure-as-a-service (IaaS). Providers, too, must pursue the right kinds of strategic partnerships, i.e., the ones that enable customers to fully exploit the many advantages of a simple, scalable and secure cloud.

Automated processes and a zero downtime architecture facilitate reliable, uninterrupted operations even when maintenance work is required or central components need to be updated. Under the hood, this works because the computing/storage resources and network infrastructure are actually kept entirely separate from the IT services and workloads, even though all of the components are managed by a common automation framework. These interfaces are the technical resources that make it simple to balance workloads between the physical locations both online and offline. The customer front-end uses integrated customer self service portals that provide automated features and applications. Acting as a hub for public and private cloud services from arbitrary providers, they work to enable a fully-featured service and hybrid cloud management product. This means customers can configure, manage and pay for any of the IaaS services offered, i.e., to suit their various needs and without having to host any of the technology themselves.

6.2.3   A High-Availability Duo: The Twin-Core Model

The standards set for cloud infrastructure are high – and are even higher when this is sourced from a single provider. Apart from cost effectiveness, top-priority items include reliability, data integrity, data protection, unlimited scalability, and the availability of data and systems. At the same time, sustainability issues are becoming increasingly important for data centers as well, due to significant increases in energy costs. Such cloud services must therefore be hosted by modern, efficient data centers with failover protection. After all, cloud-served company infrastructure and business processes only add value if the technology is guaranteed to run smoothly. This is where a "belt-and-braces" approach proves its worth, and "twin-core" data centers work according to this very principle. All of the data and systems are mirrored continuously to a twin data center, where they remain available even if one of these "twins" suffers faults and outages – due to extreme weather conditions or flooding, for example. Germany's largest data center in Biere, Saxony-Anhalt forms just such a twin-core with the data center in Magdeburg. Thanks to this arrangement, the 40,000 square meters of data center space (once fully complete) will offer availability of up to 99.999 percent, which translates into a maximum downtime of around five minutes for a whole year. Comprehensive security models, which include the deployment of IT security technology as well as numerous physical security precautions – such as access control points and surveillance systems – also keep unauthorized individuals far away from corporate data and infrastructure. The architecture of such a modern data center is also able to meet increasingly stringent requirements concerning energy costs: Paired with high-density deployment of installed IT systems, the TwinCore concept cuts operating costs and CO_2 emissions alike. Although the data centers need less space and fewer servers, they simultaneously keep IT performance high thanks to the deployment of standardized cloud platforms – which in turn cuts the costs of cloud hosting for the customer.

6.2.4 All a Question of Standards

To better understand guarantees made for high availability, we should first make it clear that systems kept available virtually around the clock are not necessarily the go-to standard. For data centers, a set of quality levels is used, as defined by the "tier" standard. This ranges from Tier 1 (low level of failover protection) to Tier 4 (maximum failover protection).

With a private cloud, data is not transmitted over the Internet but over a secure network and is hosted by high-efficiency data centers whose location is actually known – in contrast to public cloud products – and which can prove to be very useful when dealing with data protection law. Suitable data centers include those whose levels of annual availability meet the Tier 3 (99.98 percent) or Tier 4 (high availability, >99.999 percent) data center standards. Such data centers are designed from the ground up for the hosting of mission-critical business applications and the storage of sensitive data. The same levels also apply for hybrid clouds. Public clouds are typically hosted by Tier 2 or 3 data centers, however. Tier 2 offers only simple redundancy, and access is not usually handled over a dedicated, secure network but is run over the Internet, as mentioned at the outset. After registering with the cloud provider, users have immediate "one-click" access to the services. Since the provider doesn't usually include individual customer support in the package, however, this variant is typically used for test systems rather than complex systems – and is certainly never deployed for business-critical corporate infrastructure.

6.2.5 "Universal" IT Security Needs a Strategy

The fact that tier standards and strategies for high availability are hugely relevant – and in fact essential for ensuring a company's ability to do business – is underlined by the major challenges facing IT and the people who work in the field. Hundreds of thousands of new viruses, worms and trojans appear each and every day. Experts from T-Systems' Computer Emergency Response Team (CERT) also report that attackers are becoming more professional and using increasingly sophisticated methods. Modern data espionage is typically motivated by financial interests. For this reason if no other, the threat is more real and more serious than ever, and it means that cyber security services form an integral element of trust in a networked world. Whether in the in-house data center or in the cloud operated by a service provider: Without needs-based, end-to-end security, companies will rapidly face problems of an existential nature.

To date, however, the security architecture of many companies has resembled a medieval city wall: a hard shell, but a soft underbelly. Perimeter security is used to corral individual IT systems, while entrances and exits are more or less well-protected. Keeping things secure while achieving increased digitalization requires more intelligent and comprehensive models, however, embedded in overall corporate strategy, as well as the continuous development and refinement of these models. Companies need to transform their IT so that they can enjoy demand-driven utilization of cloud-based resources (for example) without compromising corporate security. Corporate Governance and Corporate Risk Management must both be involved here. In addition, the topic of data protection must also be given top priority: Germany has a clear geopolitical advantage here that companies should make the most of. Not least because the Federal Data Protection Act is one of the toughest laws governing informational self-determination, and the use of personal data requires either an explicit legal justification or the consent of the person affected.

Accordingly, data integrity and data protection aren't simply checklist items to be worked through: Corporate security requires a high-quality strategy and this, in turn, requires a tailor-made security assessment (see Advanced Cyber Defense, for example). This not only provides a set of results that can be used to derive a workable security strategy but also yields reports about security incidents that must be dealt with in accordance with the corporate security strategy. Specialists and specialized units who focus solely on security work can also help to raise awareness of attack vectors and scenarios, and ensure that potential hazards specific to certain solutions can be identified at an early enough stage.

6.2.6 The IT Quality Factor, or the Pipe Dream of Perfect IT

Even if the technologies themselves meet the highest standards, companies still need to be clear about one thing: There's no such thing as perfect IT. Not in the real world, anyway. Wherever IT is used, there is always the possibility that a system outage will occur. And there are plenty of reasons: a technical defect in installed components, breakdowns caused by natural disasters or bad weather, or simple human error. Nor is it simply enough to use technology as a precaution against potential outages – with high-availability, redundant infrastructure models, for example. Facilities also need to be in place for resolving faults rapidly and systematically in cases when a system decides to go off the rails. Downtime is expensive, after all: as much as 5,600 US dollars a minute according to some accounts (cf. Lerner 2014). This is an area where programs for securing IT quality, whose strategies span all levels of the company, are becoming more popular. One example would be appropriate change and incident management plus clearly defined, well-established processes that are regularly audited and measured in KPIs. Not least because IT availability isn't just a question of quality in a techno-logical sense: It is also a question of quality for the organization itself and the staff who works there ("processes, people and platforms"). And this high standard of quality must also be valid for the cloud provider's partners. Only with appropriate supplier certification can customers benefit from true, end-to-end IT availability, which not only keeps the risk of faults to a minimum in the first place – by the use of appropriate preventive measures – but also guarantees the fastest possible resolution of such faults by using a set of clearly defined and contractually agreed processes.

6.2.7 Greater Efficiency Under the Hood

Companies benefit from the availability of shared resource usage in the cloud and the fact that these services are provided dynamically according to demand. This enables the implementation of flexible business models that themselves facilitate the accelerated roll-out of scalable services, independently of both geographical location and company size. Any company's decision to adopt a cloud solution is also materially influenced by the factor of cost optimization, however. Accordingly, it is important that every process runs with maximum efficiency under the hood. Five levers, whose effects mesh together like gearwheels, are crucial here: standardized services, consolidated systems, automated production, global supply capability and optimized capacity utilization. With in-house solutions in particular, it is often difficult to standardize across a complete portfolio of products. Yet this is precisely where working with a service provider brings further advantages – always assuming full use of the latter's options in the service of the customer. For the greater the degree of standardization in a provider data center, the larger the ultimate benefit for companies, even at later stages. If standards are implemented, consolidation onto just a few systems is possible. This, in turn, sets the stage for optimum automation, since automating a handful of large systems rather than many small ones is considerably more straightforward. In large environments, many thousands of systems can be consolidated and automated simultaneously. This not only saves physical space but ensures better capacity utilization, just as with global services and support structures, since they are uniform the world over and not tied to any one location. Infrastructure utilization can therefore be optimized and loads can be distributed dynamically. This, in turn, initiates a positive chain reaction: The costs for factory downtime are minimized, which results in lower unit costs, which in turn generates more favorable conditions for the customer.

In this context, economies of scale have a decisive role to play, since comparable unit costs are effectively out of reach for in-house operations. After all, few companies are capable of operating the perfect combination of standardized technology and the necessary degree of process automation entirely in-house at a level of efficiency that can also match the data center's benefits of scale in terms of energy procurement. Such benefits only materialize following consistent implementation of the underlying models, paired with a large customer base. Close-meshed monitoring is also important: Only continuous measurement of the results by means of KPIs (utilization, costs, etc.) creates a basis for their further optimization. Ultimately, broad-based efficiency is therefore the essential foundation for a cost-optimized cloud. Even then, its tremendous economic potential has still to be fully exploited: As stated at the outset, the cloud-accelerated economy has yet to reach its top speed.

6.2.8   Outlook: The Software Defined Data Center

The logical next step in efficiency is to virtualize the infrastructure in its entirety, and here we are seeing a new trend with a bright future: the Software Defined Data Center (SDDC). Constant growth in data volumes, combined with the routine appearance of new requirements such as realtime reporting on information, is placing greater and greater demands on IT infrastructure. According to one report, the global market for SDDC is set to reach a staggering 77.2 billion US dollars by 2020 – equivalent to an average annual growth rate of 28.8 percent (cf. Markets and Markets 2015). In Europe, the annual growth rate forecast is as high as 40 percent. So what does SDDC bring to the table? SDDC offers a centralized point of control for reducing IT infrastructure complexity and simplifying its management. Hardware such as storage, CPUs, firewalls, routers and switches is virtualized and can be configured in the SDDC model without physical access to infrastructure components. The end result is a fully "elastic," demand-driven allocation of resources that is both automated and scalable. In a nutshell, this means CIOs effectively get a wall socket for the application, which then simply needs to be integrated into their business processes. Operational processes are handled by the provider, who moves the application workload between the private and public cloud as needed – and naturally not before implementing appropriate levels of encryption to fulfill data protection requirements.

6.2.9   Conclusion

There's no such thing as "the" cloud. Just as each company is different, so too are the individual requirements for corporate IT and hence cloud services. These services are provided by a number of cloud models, however. It is important that businesses choose a strategy that works for them and sign up for a cloud service that offers enough flexibility to adjust this strategy to meet new requirements whenever this may become necessary. Flexibility is, after all, the key benefit of the cloud. Always assuming that the infrastructure is operated by a service provider who can offer the necessary platforms and platform strategies. Efficient, reliable and cost-effective cloud operations require more than just a few server racks in the in-house data center: They require experience, technology and quality. Ideally, the data center will also be located in Germany to ensure adequate provisions for data protection are made based on the Federal Data Protection Act. This overall package is also essential if the cloud is to make good on its promise of equipping companies to face the future. Technical specifications are one thing, but innovation depends on much more than a virtualized environment.

References

Behrens, F. (2015). *Neue Wege im Automarketing: Internet & virtuelle Showrooms machen dem Autohaus zu schaffen*. In: kress.de. Accessed August 25, 2015, from https://kress.de/tagesdienst/detail/beitrag/132126-neue-wege-im-automarketing-internet-virtuelle-showrooms-machen-dem-autohaus-zu-schaffen.html

Binder, E., & Mortsiefer, H. (2014). *Virtuell Gas geben*. In: tagesspiegel.de. Accessed August 12, 2015, from http://www.tagesspiegel.de/berlin/audi-city-am-kurfuerstendamm-virtuell-gas-geben/9432044.html

Crisp Research (2015). *Digital business readiness*. Authors: René Büst, Maximilian Hille, Julia Schestakow. Accessed August 12, 2015, from http://www.crisp-research.com/report/digital-business-readiness-wie-deutsche-unternehmen-die-digitale-transformation-angehen/

Gabler, T. (2015). *BMW verspricht die Revolution der Mobilität*. In: internetworld.de. Accessed August 25, 2015, from http://www.internetworld.de/technik/smart-wearables/bmw-verspricht-revolution-mobilitaet-881300.html

Gaudin, S. (2014). *Hybrid cloud adoption set for a big boost in 2015*. In: computerworld.com. Accessed August 26, 2015, from http://www.computerworld.com/article/2860980/hybrid-cloud-adoption-set-for-a-big-boost-in-2015.html

Lerner, A. (2014). *The cost of downtime*. In: Gartner Blog Network. Accessed August 19, 2015, from http://blogs.gartner.com/andrew-lerner/2014/07/16/the-cost-of-downtime/

Markets & Markets (2015). *Software Defined Data Center (SDDC) Market by Solution (SDN, SDC, SDS & Application), by End User (Cloud Providers, Telecommunication Service Providers and Enterprises) and by Regions (NA, Europe, APAC, MEA and LA)—Global Forecast to 2020*. Accessed August 25, 2015, from http://www.marketsandmarkets.com/Market-Reports/software-defined-data-center-sddc-market-1025.html

PWC (2015). *Report commissioned by JDA software. The omni-channel fulfillment imperative*. All accessed August 18, 2015, from download link: http://now.jda.com/ceo2015.html. Individual sections also available from TecChannel.de: http://www.tecchannel.de/ecommerce/know-how-und-praxis/3199822/die_digitalisierung_fordert_den_handel/index3.html

T-Systems MMS (2015). *"Die Transformation definiere ich als eine Reise, die man—spätestens jetzt—beginnen muss"; Interview mit IT-Chef und Innovationstreiber Pascal Morgan, Coca-Cola*. Accessed August 24, 2015, from http://wegweisend-digital.t-systems-mms.com/interviews/pascal-morgan-das-optimum-der-digitalen-transformation.html?wt_mc=osm_3:15:15

Zukunftsinstitut (2015). *Retail Report 2016*. Authors: Janine Seitz, Theresa Schleicher, Jana Ehret; Managing editor: Thomas Huber; Published by: Zukunftsinstitut GmbH.

Frank Strecker Senior Vice President Cloud Partner Products & Ecosystems, is responsible for global cloud computing and partner sales business at T-Systems International GmbH. His work focuses on the development and expansion of these strategic business segments while ensuring the integration of all units in terms of cloud computing and partner sales. A member of the Leadership Team at Deutsche Telekom AG, he also heads the company's Cloud Leadership Team. With more than 17 years of experience in the ICT industry, he has held numerous national and international management positions at IBM. Frank Strecker holds a degree in technical management from the University of Stuttgart, Germany.

Jörn Kellermann SVP Global IT Operations at T-Systems International GmbH, has global cloud and DC operations responsibility for the company's entire customer base. This includes provisioning and operational duties covering both hardware (IP networks and data centers) and software (SAP, messaging and custom applications). Jörn Kellermann has worked in IT for over 20 years. After graduating in computer science and business management and following a period as an independent consultant, he joined systems provider debis (now T-Systems) in 1999. Since then, he has held a number of positions in the company's Sales, Consulting and IT Service Provision units. He previously headed the global Dynamic Platform Services (DPS) division.

No Innovation Without Quality

7

Anne Teague

Like many food and drink companies, Heineken's top IT priority is to tackle big data. It will form the foundation for all of our digital innovation and support our strategy of developing a deeper understanding of the customer. Heineken is a consumer brand, but its activities are predominantly selling business-to-business, so we have typically been one step removed from the consumer. As a result, it has been difficult to truly understand how, when and where our beers are drunk. But now that is changing.

There is a huge – potentially overwhelming – amount of data becoming available. That data flows at great speed, from a huge variety of sources, in many different formats. Our ability to innovate will depend on how quickly we can analyze this data and convert it into actionable business intelligence. Throughout the process, we must ensure that the data is secure and privacy is protected. If we can do all of this successfully, we'll be able to improve the customer experience and the products, and reduce our environmental footprint too.

7.1 Social Media Drives Data Growth

One reason for the growth in data is social media, a powerful tool for brands to connect directly with their consumers, bypassing the supply chain. Heineken's brand Don Equis was the first to achieve a million Facebook fans and now has 2.9 million. That's just one of more than 250 brands we own that are growing their fan bases. Outside of these direct relationships, we also need to be able to listen to what people are saying about our brand online. A reputation is won over years but can be lost in hours. Complaints can quickly go viral, sometimes based on erroneous information, and we have learned it is important to be able to identify and correct any misunderstandings promptly.

A. Teague (✉)
Den Hague, Netherlands

© Springer International Publishing Switzerland 2017
F. Abolhassan (ed.), *The Drivers of Digital Transformation*, Management for Professionals, DOI 10.1007/978-3-319-31824-0_7

7.2 Gathering Data from the Internet of Things

Another driver for the growth in data is the Internet of Things. According to analyst
house Gartner (2014), there will be 25 billion devices connected to the Internet by
2020. Deployment is still in the early stages in the food and drink sector, but
connected devices have the potential to improve both the customer experience
and the quality of information available to the manufacturer. For example, smart
kegs could be used to monitor the freshness of beer in bars, ensuring a premium
experience for customers and also providing the manufacturer with information on
consumption. In retail outlets, fridges could recognize customers and offer relevant
manufacturer promotions to them when they're selecting their drinks.

Devices in the home could also be connected. The Heineken Sub is a home
draught beer machine available in France, Italy, Spain and the Netherlands. In the
future, phone apps linked to the Sub could be used to show customers when they
need to reorder and deliver relevant promotions to them too.

7.3 Realizing the Value of Big Data

As more and more devices become connected, we'll be able to gather data from
production to consumption: Learning more about our manufacturing and transportation processes, and how our products are bought and enjoyed. We've already
begun to fill the big data pool, but this information will become most valuable when
we are able to mix data from multiple sources, analyze it and spot previously
unknown correlations and relationships. This capability will be fundamental for
enabling digital innovation and realizing value from it.

7.4 The Constraints on Digital Innovation

All too often, digital innovation is not empowered but *constrained* by the IT
systems an organization has. Large companies typically have a patchwork of legacy
systems built up over decades and through acquisitions, making the IT inflexible
and hard to manage. In many cases, it is ill-suited to the big data world where
information needs to be able to flow freely between applications and be analyzed in
multiple contexts.

Integrating and launching a new application could take 18–24 months. It's hard
to reconcile that with the need to be an agile organization responding to market
opportunities quickly. At Heineken, for example, we have projects that we know
about a long time in advance (such as our sponsorship of the Champions League
and the James Bond films), but our local marketing teams also need to respond to
opportunities that arise in their markets at short notice. The entire lifecycle for a
new digital product might be just three months, from inception to retirement. IT
organizations must reinvent themselves so they can deliver digital innovation in
those timeframes.

7.5   Reinventing the IT Function

To achieve this transformation, there needs to be a shift in the IT team's mind-set. If the team is spending most of its time managing the plumbing of the IT (as many do), then there is little time left to support the business with these new applications. The goal should be to establish a high-quality system that requires minimal day-to-day intervention from the in-house team, so that they are free to work creatively with the rest of the business. By definition, innovation is about doing something for the first time. It's risky for the business owners, so it's essential that they feel they can depend on the IT team to support them. This is especially important in a federated organization such as ours, where the centralized global IT function must win the trust of the operating companies worldwide and be seen as the best partner for delivering innovation. If the IT team is tied up in maintenance, it will be hard for them to give the business users the time and attention they deserve.

Heineken outsources the management of most of its infrastructure to T-Systems Dynamic Services, which takes care of hosting, operations, technical application management, maintenance, administration and security. Because T-Systems manages the operational infrastructure layer, the global IT team at Heineken is able to focus on more strategic initiatives in partnership with the business.

It won't be easy for the IT team to pivot from plumbing to innovation. The technological change and outsourcing process take some time, and it is important not to underestimate the cultural change required too. Some IT team members might initially feel that their job is threatened when external partners are brought in, or feel that their authority is being eroded within the business. Clear communication and training will be essential to ensure that the team first of all understands and then benefits from the opportunities for job enrichment that this adjustment brings.

7.6   Defining the High-Quality IT System

The first step in enabling a more agile IT function is to establish a high-quality IT system that it, and the entire business, can depend upon.

Quality is a slippery concept to define, and in the past metrics such as uptime have often been used as proxies. End users have become increasingly demanding of the IT that they use at work. At home they use web services (such as Gmail or Facebook) and desktop email software. Downtime for these is practically zero. When people get to the office, they expect that same smooth experience and want to trust that the IT will always be available so they can focus on doing great work. Downtime is a blunt instrument for measuring user satisfaction, though, which is the real end goal.

At Heineken, we have abandoned key performance indicators in favor of a focus on the end user experience. In our view, a high-quality IT system is one that makes users happy, which means enabling them to do their job effectively, both day-to-day and when launching entrepreneurial projects. That requires reliability, of course, but scalability and agility are equally important, and flexibility is also key. These are the characteristics that define a high-quality IT system in the eyes of the users.

There are three facets to a high-quality IT system: the platform, the people, and the processes. All of these must be optimized to satisfy the end user's needs.

7.7   The Platform

When I assumed the CIO position in 2012, one of my top concerns was to consolidate our IT environment and make it more manageable, so we could more easily respond to the business requirements. While we still have a significant estate of legacy IT, we are aiming to move as much of our IT as possible into the cloud, and this is where our new investments are focused. As well as reliability and built-in redundancy, the cloud offers us scalability. This will become increasingly important as we embark on big data projects.

We are committed to the concept of software-as-a-service. When we chose Office365, for example, we did so knowing that the vendor would update it and add new functionality over time without us needing to worry about it. This is a pain-free way for us to keep our systems current, without distracting the IT team from more strategic ventures.

The cloud also satisfies our requirements for a more agile IT function, making it easy to spin up new services and applications. We can ramp up resources at short notice too, so it's easier to support digital campaigns with uneven or unpredictable demand, or to respond to any other business change quickly.

There is a built-in benefit to the cloud's pricing model too in that it encourages the right behavior from business functions. Under an "all you can eat" pricing model, there is no incentive to innovate in the IT, rationalize it or even retire applications effectively at the end of their life. With a focus on launching new applications, it's easy for this to be overlooked and for redundant applications to accumulate. This is especially true as the lifecycle of digital products increases, and we see more and more short-term and local campaigns launching across our territories.

With a pay-as-you-go model, or in the Heineken world "pay per drink," the business pays less if it uses less resources. This approach rewards the right behavior and helps to ensure that resources are not being wasted. We have ambitious targets to reduce our carbon emissions, and effective use of IT can make a contribution towards that. Without this incentive, users are likely to be more risk-averse and less innovative. The risk of downtime is a given in any IT project, so business leaders are unlikely to take it on if there's no possible reward.

The cloud is enabling us to move towards the dynamic workplace too. People are passionate about the devices they use at home, including phones and tablets that provide instant information through an intuitive and enjoyable user interface. At work, they increasingly expect to use similar technology. The idea that you would have to go to a specific desk or device to find some information seems antiquated. People want to be able to access corporate systems in the same way they access their personal email and social networking applications: anywhere, any time and using whichever device they have available. From a business perspective, this capability is essential for agility and competitiveness.

One way to achieve user satisfaction is to allow them to use the devices they are most comfortable with, so we are now exploring bring-your-own-device (BYOD) and bring-your-own-mobile (BYOM) strategies. There are many challenges associated with these strategies, including how you reconcile staff freedom with company expectations. Heincken is a premium brand, so it might create the wrong impression for a sales rep to make a customer visit with a beaten-up old laptop, for example. From an IT delivery point of view, though, it shouldn't matter whether the user prefers Android, Apple or Windows: Our goal is to accommodate whatever makes the user happy.

As a step towards enabling the dynamic workplace and BYOD, we have implemented two programs called HeiHosting, which consolidated our SAP instances and non-SAP software, and rationalized our server landscape. This was an important first step because if your software and server landscape is localized, it is difficult to be agile and provide consistent remote access to applications. Simplifying the IT environment was the most important project in our strategic plan.

We consider standardization to be essential for a high-quality IT platform. We want to empower business users to use off-the-shelf solutions from marquee partners, without customization, so we can deploy more quickly and at lower cost. Customization carries risks and implies a need to retain specialist knowledge on how a system has been modified. It also takes time that would be better invested in more strategic initiatives, both on the IT and business side of the project. We rely on T-Systems' roadmap and offering so we don't have to reinvent the wheel, and so we can free up time and budget for innovation.

One of our guiding principles for the platform is that we don't want to be the first to deploy a new technology. That might seem to stifle innovation, but in fact the opposite is true. For us to be innovative, we need to be agile, need to have systems that users willingly adopt, and need to have a strong relationship between the business users and the IT team. All of these could be threatened by deploying an unproven technology that resulted in downtime or other failures.

7.8   The People

Having the right platform is only part of the puzzle: Its success depends on the people who are responsible for it. We can take it as a given that the people have the necessary expertise and experience required, so what defines the quality of the people is how well they are able to work together and communicate, especially between the customer and provider organization in an outsourcing arrangement.

Internally, the IT team has an important role in coordinating the ecosystem of partners and ensuring they satisfy the business requirements. Outsourcing doesn't mean that the IT team can sit back and not do anything. Externally, outsourced partners should take responsibility for delivering the day-to-day IT quality, and for helping to evolve the platform.

To get the best out of an outsourced partner, it is important to focus on the long-term relationship. At Heineken, we are not interested in enforcing penalties when service level agreements aren't met. This might sound unusual, given that the outsourcing industry has historically relied on penalties to govern providers. We strongly believe, though, that it is short-term thinking to see a mistake and then seek punishment and compensation for it, rather than working cooperatively on a solution. Penalties don't fix the underlying technical problem and don't reassure the business that the issue is under control either. At a time when the customer and provider should be working more closely together because a problem has come to light, penalties drive a wedge between them, demotivating those who are most influential on the success of our IT. When entering into an outsourcing relationship, it's healthier to think of it like a marriage, where mistakes are forgiven, and each party supports the other in the long term.

That's not to say that the outsourcing company has an easy time. Its people need to develop their role beyond keeping the systems running and hoping the customer doesn't bug them. At Heineken, we are a beer company and not an IT company. We turn to our ecosystem of providers for advice, and expect companies such as T-Systems to proactively identify opportunities to increase standardization and introduce new technologies that will better support our requirements. In the way that the internal IT team must focus on innovation and become a trusted advisor to the business, outsourcing providers must be willing to support our entrepreneurial projects and become trusted advisors to the IT team.

This is only possible if there is honest and open communication on both sides, based on a deep understanding of each other's businesses. That means discussing not just the opportunities but also the difficulties. In an environment reliant on penalties, this kind of communication can be suffocated by the fear that it would call attention to weaknesses. In a long-term relationship, discussions like these can only strengthen the partnership and the IT systems by providing a foundation for robust long-term planning.

This kind of relationship requires a complete change in skill set. The internal IT team needs to work with outsourcing providers as partners and colleagues, and not as suppliers. This is much more challenging than just checking whether the IT is working, and working out which penalties apply if not. The external team needs to go beyond the technical skills required to manage today's IT systems and become more strategic, aligning with the customer's long-term goals and proactively helping to evolve the system towards them.

On both sides, there needs to be an unprecedented level of trust. For those who have previously worked under a penalties-led framework, it will take considerable effort to become less defensive and more open. Many roles will be transformed from operational and responsive, to strategic and proactive. It is still rare to find people who have the complete skill set required, so it is essential that customer and provider businesses both help their staff to broaden their talents.

There might be resistance to this change internally from those who would rather focus on IT plumbing than more strategic work. At Heineken and T-Systems, our communications teams have joined forces to help explain to the Heineken operating companies what the changes are and how they will be affected. Clear communications within the companies can make a big difference in helping every-body to see the opportunities that a closer partnership will bring.

7.9   The Processes

The customer and outsourced provider need to work together to plan which processes should be implemented. As a first step, there needs to be a business-IT alignment process to ensure that the technology is meeting the needs of the business owners. This ensures that the outsourcing provider offers the right solutions, and can also identify any gaps in requirements. T-Systems explains its roadmap to us, and if there is something we need that they cannot offer, we are able to work together to find other providers that can plug the gap.

It is important that the customer organization has a clear vision of the desired architecture to ensure that any changes to the systems are consistent with what the customer wants its IT to ultimately look like. Without a vision, the risk is that the architecture could gradually evolve in a direction that the customer would not ideally have chosen.

Service level agreements remain useful for managing expectations on both sides, but it's important to think of what you want the relationship to achieve, and find effective ways to track progress towards that. In our relationship with T-Systems, we measure the level of standardization and server consolidation and have an index for cost reduction. These reflect the strategic imperatives in the relationship, and show how quickly we are moving towards a platform that will better support agile innovation.

We also aim for Zero Outage in our partnership with T-Systems, which is a requirement to satisfy end users and ensure that our business can operate uninter-rupted. Heineken is increasingly dependent on automation supported by IT for our production processes, so downtime can have a significant effect on our business – and it's impossible to recapture the time lost if a production line is stopped. If a brewery can't brew any beer for a day, it would be a disaster.

T-Systems has processes to achieve zero outage that span operations for day-to-day availability, and projects, which is when the greatest risk of downtime occurs. Major incidents are prevented using effective project planning, employee certifica-tion, regular system monitoring and end-to-end business service monitoring. The processes increase the availability of Heineken's network and monitor any risks in the operating environment. The effectiveness of these processes is measured using the TRI*M index, which stands for the three Ms: measuring, monitoring and managing. To track its progress, T-Systems commissions market research firm TNS Infratest to ask customers how T-Systems scores in these areas. Over the last three years, T-Systems has been able to raise Heineken's rating from 40 to 110 as a result of focusing on high-quality platforms, processes and, especially, people.

It is also important to introduce processes that enable partners in the outsourced ecosystem to speak to each other. Instead of being the middleman in the process, the customer organization should empower its partners to cooperate with each other in the best interests of the customer and the outsourced providers.

In our case, T-Systems works closely with other ecosystem partners, speaking weekly and meeting monthly, to discuss how they can jointly prepare to meet Heineken's requirements. Top of the agenda are innovation and rationalizing the app landscape. If there are operational problems, these companies can work together to resolve them without necessarily needing to involve Heineken in that process. While our IT team will sometimes need to contribute our company perspective, if they can be excluded from more routine inquiries that further liberates them to work on digital innovation with end users. High-quality processes should, as far as possible, enable the IT systems to be managed independently of the in-house IT team.

This is only possible when everyone enters into the relationship in the spirit of partnership, and when the processes help to reinforce that. The customer needs to trust that its providers will work in its best interests. The ecosystem partners will sometimes be in competition with each other and need to communicate openly about that and turn to the customer for a judgement on how best to handle that. Everybody needs to be confident that they will be supported by their colleagues in other organizations, and that if something does go wrong, they will work together to resolve it.

To ensure the organizations remain aligned, there is also a continuous dialogue between the top managers at Heineken and T-Systems. A strong relationship can only be built on trust, with both parties sharing their goals, upcoming innovations, and aspirations for the partnership.

7.10   Conclusion

New technologies, including big data and the Internet of Things, bring huge opportunities to companies that want to better understand their customers and find new ways to delight them. In order to take advantage of these opportunities, there needs to be a high-quality IT system that offers not only reliability, but also agility, scalability and flexibility. The right platform is the basis for a high-quality IT system, but it must be supported by the right people and the right processes. In particular, the IT teams need to develop a more strategic outlook that enables them to support the business requirements for innovation, and there needs to be a trusting relationship between companies and their outsourced partners. This provides a stable basis for day-to-day and strategic management of the IT systems, and a springboard for agile innovation. There can be no digital innovation without a high-quality IT system to enable it.

Reference

Gartner (2014). *Gartner says 4.9 billion connected "things" will be in use in 2015* (Press release). Accessed August 29, 2015, from http://www.gartner.com/newsroom/id/2905717

Anne Teague has more than 20 years of experience in managing and developing global IT programs for large FMCG companies. Between February 2012 and August 2015, she was the Global CIO of Heineken International and in this capacity responsible for successfully implementing the standardization and globalization of systems, processes and data. Today, Anne Teague works as an independent consultant supporting other companies in their IT transformation processes.

The Counterculture of Silicon Valley

Steffan Heuer

It only takes a few weeks in Silicon Valley to gather ideas and establish long-lasting relationships – with some well-calculated culture shock included. But what exactly do German executives take home with them after visiting this epicenter of global innovation?

Thomas Neubert can open the doors to big companies or small startups in Silicon Valley – but regardless of which it is, his visitors from Germany usually have three goals: to get inspiration, sound out partnerships and maybe even lay the groundwork for investments. Neubert has been head of the Business Development and Partnering Group for Deutsche Telekom in Mountain View since 2012, and after spending a total of 25 years in the region, he knows this high-tech stronghold better than anyone. This benefits his guests, for whom Neubert regularly organizes what are usually week-long "innovation expeditions" in the Valley.

A week sounds like a short time, but these expeditions are preceded by months of preparation which involve identifying "the most important key players, venture capitalists and exciting startups in each technology or sector." If nothing else, this helps channel his visitors' curiosity and interests so that Neubert's European colleagues and American partners get the maximum benefit from such short trips – by jointly spotting upcoming trends early on, testing their potential market rollout in Europe and entering into partnerships and investments that pay off for both sides.

Take Deutsche Telekom, for example: By the summer of 2015, around 80 visitors had experienced this intensive exposure to California's high-tech center – everyone from board of management members to the marketing heads of all of Deutsche Telekom's European affiliates. "We've established a credible presence in the Valley which makes it possible for us to spot trends and ideas over here and, conversely, to share our own ideas," Neubert says. "This exchange

S. Heuer (✉)
Heuer Media, 1266 De Haro Street, San Francisco, CA 94107, USA
e-mail: Steffan@heuermedia.com

© Springer International Publishing Switzerland 2017   83
F. Abolhassan (ed.), *The Drivers of Digital Transformation*, Management for
Professionals, DOI 10.1007/978-3-319-31824-0_8

regularly gives rise to partnerships for the European market which are very valuable to any startup from the Valley wanting to grow quickly."

8.1 Endless Tech Boom

Deutsche Telekom is not alone in its desire for fast, dynamic access to the epicenter of the networked world. European entrepreneurs and managers have long been drawn to the West Coast. Spurred on by the relentless tech boom, more and more German companies are dispatching employees to the region between San Jose and San Francisco.

They often spend just a few weeks or months in northern California, but there is a definite trend towards staying longer – sometimes for several years – so they can scout for trends permanently instead of just intermittently, get closer to the deal flow of local venture capitalists and generally be inspired by the faster, more open culture of the whole Bay Area. In extreme cases, some German entrepreneurs are spending more time in California than at their headquarters in Europe, because after a few years of continual expedition trips they have realized that their international customers and their Californian partners and investors want them to be more present – and their expanded presence in Silicon Valley is a good selling point back in the Old World.

Technical innovation happens outside of Silicon Valley too, of course, but hardly any other region has produced as many success stories since the 1950s. Semiconductor production was followed by the computer industry and then – since the triumphal march of the web – by the Internet economy that has grown up around applications for connected living and the Internet of Things. "The Valley" first gave birth to Intel, Cisco and Apple, and later to the new generation of globally successfully companies such as Google, Facebook and Twitter.

8.2 Magnet for Capital and Creativity

The statistics paint a clear picture. The flood of innovations shows no sign of letting up because local startups are continually the source of important (and sometimes controversial) sparks for the digital transformation of every sector of industry. In 2014 alone, 25.3 billion dollars of venture capital flowed into the region, which now has a good three million inhabitants. "In Silicon Valley, the idea gets financed, not the success," wrote Ulrich Grillo, President of the Federation of German Industries (BDI), in the *Handelsblatt* newspaper, pointing out what is perhaps one of the most important cultural differences. In terms of economic power, American startups pull in seven times as much risk capital as their German counterparts. The German startup world is lagging behind especially as regards financing opportunities, says Grillo. The large number of startups contributed significantly to the fact that, in the past year, nearly one out of every ten companies that went public in the USA was from Silicon Valley.

It is no wonder that German companies want to keep a closer eye on this creative activity which is attracting smart people from all over the world. Neubert's team at Deutsche Telekom therefore evaluates dozens of startups working in areas such as big data, the Internet of Things, the connected home, wearables and consumer services. The "expeditions" then pack in four or more meetings every day, where large companies such as Cisco or Apple outline their vision for the networked future and startups present solutions which are sometimes only half-formed. "These visits always have a clearly defined agenda," Neubert says, "if for no other reason than to avoid 'scattergun' innovation tourism." Above all, many of these meetings offer an opportunity to discuss new ideas or potentially profitable areas of business much more openly than is usually the case in meetings or sales pitches for the purposes of customer relationship management.

8.3   Two-Pagers Instead of Hundred-Page Contracts

Other German companies are taking an equally targeted approach. Stefanie Kemp, head of IT Governance at RWE and a member of the RWE Innovation Hub leadership team, has been traveling back and forth to Silicon Valley for a decade "to see where we can reasonably introduce innovations into our business segments which are stable and have growth potential." Major trends such as digital transformation, Industry 4.0 and the Internet of Things play an especially important role for Kemp. For a company like RWE, these trends are reflected in activities such as decentralized energy management and solutions for smart and connected living.

Kemp mentions Enlighted as an example of a startup whose ideas have effectively "electrified" her company. This spinoff from Massachusetts Institute of Technology (MIT) specializes in sensor-controlled room lighting which responds to human movements. C3 Energy, another example, develops smart grid applications and views energy supply as an aspect of the sharing economy. Among other things, this entails flexible and intelligent billing and management platforms which could make it possible for potentially millions of households with solar collectors to share charging stations for powering electric cars, for instance. "Every day the Valley produces new solutions, products and innovations, and contracts are signed for them immediately. It all happens not with patent applications or hundred-page agreements, but on the basis of two-pagers, here and now – it definitely has its own culture," Kemp says.

8.4   Less Complexity – More Risk

It is in this sense, above all, that RWE managers take away a better understanding of innovation culture from their visits. Kemp is always asking the same question: "How does Silicon Valley, as its own community, manage to keep the innovation cycle going?" Sometimes it has been "simple realizations that we've taken home with us: We're also working to change our corporate culture. It comes down to reducing complexity and establishing a new culture of error and risk. We want to embrace the idea that failure can also be a success as long as you learn from it." In order to become part of the ecosystem in California and connect with potential partners more quickly, RWE has had a five-member team on the ground since March 2015.

The energy company E.ON has similar plans for Silicon Valley. "Nowhere else in the world has the same confluence of creativity, capital and entrepreneurship. It's is a unique breeding ground for innovation and speed, and it's what makes the place such an effective incubator for ideas," says E.ON CIO Edgar Aschenbrenner. The E.ON group is pursuing three specific overriding goals in Silicon Valley: "First, we visit companies that we believe can contribute to the further development and implementation of our digitalization strategy. Second, we gather cultural food for thought, such as how to use social media to make the hierarchies and structures in the company more permeable to ideas." For example, E.ON now has an "Executive Hub," which is like an internal Facebook for the board of management and top-level managers. And third, E.ON picks up ideas in California for how it can offer new, web-based services to customers. "We look to see how you build something like that architecturally, keep it simple to use and underpin it with a robust IT security and data protection concept. Then we implement it at home with our own IT partners," Aschenbrenner says. One concrete result of this was the creation of a Big Data Lab by the Technology & Innovation unit at E.ON which can predict the failure of assets in order to initiate proactive maintenance, for example, or can analyze consumption figures for business customers in order to develop savings ideas and efficiency measures. "The lab employees have gone off the beaten path and can now deliver results within just a few days, or even hours," says Susana Quintana-Plaza, Senior Vice President Technology & Innovation at E.ON.

In the summer of 2014, E.ON Technology & Innovation also set up a branch office in San Francisco which acts as a "permanent bridgehead" to new technologies, processes, issues and companies. The office is run by Konrad August-in, who is responsible for strategic co-investments. Prior to this, the company spent two years sending employees from its Venture Team to Silicon Valley every six months and putting them up in the office of a local venture capital partner.

8.5   Tangible Results and a Dose of Euphoria

"How long you stay depends on what you're trying to do," Augustin says regarding the optimal timing. "If it's more about business, then its good to spend a few days or weeks gathering impressions and experiencing the different way of thinking and the willingness to take risks. But if it's about investment, you need more time to form networks and cultivate dialogue." Augustin describes his job as "market development," which involves contacting young companies who have moved beyond the concept-testing phase as well as venture capitalists. "Scouting is important, but it's always just the first step." By mid-2015, E.ON had invested in seven startups from

Silicon Valley (out of a total of 12 in the USA and Europe) whose solutions take account of the coming changes in the energy market.

But in Edgar Aschenbrenner's experience, it's important to not "just look at startups. We naturally go to Cisco and Intel, too, to find out what the next-generation chip will look like and where computers are headed. When you see how much money and manpower these companies are investing in research and development, you realize immediately that they're the powerhouses."

It is no accident that Silicon Valley offers such an unbelievably dynamic – and, in the eyes of foreign visitors and businesspeople – very attractive mixture of large companies with a global presence and a vast ecosystem of small startups working in nearly every imaginable niche. In fact, it is the product of a historical development that can be traced back to before the birth of Moore's Law. Just as a reminder: In a short essay written for Electronics Magazine in 1965, Intel co-founder Gordon Moore said that the complexity (and thus the power) of computer chips doubled every 12 months. Ten years later he revised his prediction of exponential growth to 24 months, and this cycle remains the driving force behind innovation in the Valley to this day. Technology consultant Rob Enderle calls Moore's Law the "heartbeat" of the region. "It has driven the Valley forward at an unprecedented pace and secured its leading global position," Enderle says.

Historian Leslie Berlin, who maintains the Silicon Valley Archives at Stanford University, thinks a total of three forces have turned the region into a Mecca for foreign businesspeople, engineers and programmers: technology, culture and capital. She says transistors were the grain of sand around which the pearl of Silicon Valley formed and continues to grow – from the first semiconductors, to the cloud and mobile apps.

Equally important is the unique culture of the Valley, which has been transformed from an agricultural region with fruit orchards into a community of inquisitive minds and inventors. The ongoing population boom has also ensured a regular inflow of fresh blood and fresh ideas which the long-term visitors from Germany can discuss and vie for with US citizens, other Europeans, Chinese or Indians. The population of the Valley tripled to one million between 1950 and 1970. In purely mathematical terms, according to Berlin, this equates to one new resident every 15 minutes over the course of 20 years. Today, one out of every three inhabitants of the region comes from outside the USA, and among college graduates the number of foreigners rises to more than 60 percent.

One thing is certain: Anyone who comes here and stays must be open to the Valley's unfamiliar structures and its (often equally unfamiliar) straightforward conversational style, which differs from traditional European norms and even from the standards of the East Coast of America. Ultimately, this nearly institutionalized love of experimentation attracts a particular class of managers and makers who are willing to accept that the Valley is different. And this difference is based, to a certain extent, on the "counterculture" of San Francisco, which has – at least in part – influenced the thinking of the local technological elite.

The third important element that makes the Valley such a huge magnet for European businesspeople and inventors is its network of large, institutional and small investors who are willing to bank on "the next big thing" even if it falls outside of the Valley's traditional core competencies of pure hardware and software. The best example of this is the rise of highly disruptive startups such as Uber or Airbnb, each of which, in its own way, has rocked or even permanently taken down an established industry, such as transportation or accommodation. Incidentally, another of the Valley's draws is its average annual income of 116,000 US dollars which is nearly twice the national average.

As many members of the "innovation expeditions" find out first hand, such "disruptions" usually only come about because established entrepreneurs or venture capitalists have created and continuously nurtured close-knit networks of young founders so that they can identify trends early on and invest in them. California labor law – which makes it possible to switch from one company to another without penalty, sometimes in the space of a day – also encourages this continual process of renewal, through which ideas and/or their originators are chosen on the basis of their utility and feasibility, round after round.

Berlin, the historian, quotes the semiconductor veteran Robert Noyce, who compared the Valley's generational changes to "re-stocking the stream I fished from." In other words, as soon as founders achieve success, they give back to the community that made their ascent possible, either through venture capital or through advice and support. Here, too, foreigners play a key role. More than half of all companies that were created in the Valley between 1995 and 2005 have at least one foreigner as a founder. "Silicon Valley," Berlin says, "is built and sustained by immigrants. No place else – including Silicon Valley itself in its 2015 incarnation – could ever reproduce the unique concoction of academic research, technology, countercultural ideals and a California-specific type of Gold Rush reputation that attracts people with a high tolerance for risk and very little to lose."

So, what exactly do short- and long-term visitors from Germany take back home with them, aside from impressions and anecdotes?

For Telekom manager Neubert, besides the prospect of future partnerships there is "always a hefty dose of euphoria." Anyone who spends a fast-paced week visiting technology giants like Apple, Cisco or Google – while meeting with a select group of exciting startups to boot – can easily feel dizzied by so much self-confident innovative power. The new ideas have to be digested and, above all, reconciled with the business operations that the visitors are accustomed to back home. Neubert's visitors therefore generally need "one to two weeks before they settle back into their daily routine and see what they can do in their own company."

During their visits, German companies also learn that everyday life in California casts a very different light on German corporate culture, in which everything follows orderly lines. "You realize that you have to be willing to test more new ideas and think in a more application-oriented way. And this," Augustin says, "is a very different way of thinking for a large company which has always defined centralized solutions down to the smallest detail. If you have your eye on decentralized solutions for end customers, you won't and can't always have an answer to every single question. If you always hesitate to practically implement and integrate changes, a lot of good ideas will pass you by."

The networking and learning opportunities in the Valley range from large symposia and conferences held by big companies and ambitious startups, to top-class discussions at the more than 30 universities and colleges in the unparalleled higher-education landscape led by Stanford and Berkeley. But co-working spaces and smaller, more casual, sometimes even spontaneously organized meetings are also an ideal arena for presenting new concepts and floating ideas to a knowledgeable audience – always in the hopes of receiving honest feedback and forging new bonds with like-minded technology and business experts. Visitors are quickly tempted to attend a lot of events where they can meet founders, established players and investors – and half a dozen such events are held every single day.

For all of the excitement, a healthy dose of skepticism is called for as well, as Sven Paukstadt notes. Paukstadt, the Senior Manager Partnering at Deutsche Telekom's headquarters in Bonn, visited Neubert's office in Silicon Valley from February to April 2013. He was immediately struck by the pronounced self-confidence of the tech scene there. "Everyone looks to the Valley, everyone thinks it's cool, and the locals have internalized that," Paukstadt recalls. "The people there are totally sure of themselves and act as though nothing that anyone else is doing matters."

After three months full of networking events and talks with potential partners, however, Paukstadt realized that German companies had no reason to hide from California's technology stronghold. "European enterprises are nowhere near as slow and sedate as they're often made out to be. And despite all the hype, a lot of startups aren't doing anything special," the Telekom manager says in retrospect. "If you've structured your processes well, you can work just as quickly and usually even more thoroughly than a startup from the Valley, especially in terms of contracts and implementation." This unexpected insight into the qualities of our homegrown ecosystem is another valuable takeaway from these innovation expeditions to the other side of the world.

References

2015 Silicon Valley Index. Silicon Valley Institute for Regional Studies. Accessed October 05, 2015, from http://siliconvalleyindicators.org/pdf/index2015.pdf

Berlin, L. *Why silicon valley will continue to rule.* Accessed October 05, 2015, from https://medium.com/backchannel/why-silicon-valley-will-continue-to-rule-c0cbb441e22f

Handelsblatt issued dated 08/28/2015; Wege zur digitalen Republik. Accessed October 05, 2015, from http://www.handelsblatt.com/my/politik/deutschland/analyse-von-bdi-praesident-ulrich-grillo-wege-zur-digitalen-republik/12245950.html

PwC PricewaterhouseCoopers National Venture Capital Association (2015, February). *MoneyTree™ Report 2014.* Accessed October 05, 2015, from https://www.pwcmoneytree.com/Reports/FullArchive/National_2014-4.pdf

Wadhwa, V. (2008, November). *Foreign-born entrepreneurs: An underestimated American resource*. Kauffman Foundation. Accessed October 05, 2015, from http://www.kauffman.org/what-we-do/articles/2008/11/foreignborn-entrepreneurs-an-underestimated-american-resource

Steffan Heuer is the US correspondent for the business magazine brand eins. He has spent nearly two decades in Silicon Valley reporting on innovation and technology at the interface of business, society and culture for leading European and American media outlets. Heuer studied history and economics in Berlin and New Orleans, and he has a degree from the Graduate School of Journalism at Columbia University in New York. His reports and analyses have appeared in The Economist and the MIT Technology Review, among other publications. He co-authored the book *Fake It! Your Guide to Digital Self-Defense* with the Danish journalist Pernille Tranberg. Heuer lives with his family in San Francisco.

China as the Frontrunner in Digitalization

9

Clas Neumann

9.1   Introduction

If we take a moment to consider how digitalization is affecting industry or our own personal lives, the vast majority of us would probably think of the self-driving car, smart homes, the factory of the future or – more generally – the Internet of Things and its parallels in Industry 4.0.

If prompted to clarify the source or key driver for such trends, all experts then immediately point to the West Coast of the USA. Most would include the East Coast, with many also citing Germany and Japan, and a few suggesting Israel or South Korea. The world's most populous and economically dynamic countries – China and India – do not usually feature in these kinds of global rankings.

While China's economic clout has earned it the accolade of being a key market and dominant producer of any digital product one cares to mention, China's prevailing image in the mind of the typical Western corporate executive is still that of a contract manufacturer. Terms such as innovation, leadership or creativity are less typically used to describe China than qualities such as scalability, efficiency, replication and simplification.

Yet it's easy to overlook the fact that China is already a leader in many industrial manufacturing sectors and not merely as regards production capacity or market size but in terms of China's ability to develop innovative products in these areas. One example worth mentioning here is telecommunications equipment, where Huawei and ZTE have not only become global market leaders but continue to surprise their customers with novel and innovative ideas – such as Huawei's VTA (Virtual Teller Agent) and VTC (Virtual Teller Center). For mobile devices, too, many new players such as Lenovo, Huawei and Xiaomi have appeared, who have made it

C. Neumann (✉)
SAP Labs India Pvt. Ltd., 138, Export Promotion Industrial Park, Whitefield,
Bangalore 560066, India
e-mail: Clas.Neumann@sap.com

© Springer International Publishing Switzerland 2017   91
F. Abolhassan (ed.), *The Drivers of Digital Transformation*, Management for
Professionals, DOI 10.1007/978-3-319-31824-0_9

their business to market devices that are able to satisfy the tastes (and price points) of customers worldwide. This has not been achieved by the mindless copying of existing devices that can be made more cheaply due to very high volume production, but by the canny optimization of conceptual ideas and a deliberate avoidance of unnecessary features (a topic we will return to later).

For now, however, we need to clarify what a 'digital frontrunner' means when applied to an entire country.

While identifying the leading 'digital' companies remains a very simple task, applying the same criteria from a regional perspective is a more difficult matter. Simplifying the problem slightly, a country (or a region) is considered to be a digital frontrunner if the geographic area in question produces an above-average number of innovations or innovative companies that are making significant progress in the field of digitalization. At the same time, however, the country in question must have a large number of users and the infrastructure in place to permit a high degree of market penetration for new products and services.

On this view, Silicon Valley is unquestionably one of the world's digital frontrunners, but Bangalore in South India doesn't really make the grade – since both the capital city and state have substandard infrastructure, despite fielding an above-average number of IT companies and software developers.

Other criteria that work to define a 'digital frontrunner' are:

- *Startups* that benefit from a positive ecosystem (i.e. the optimum interplay of available talent and capital) to engage in continuous research on new and pioneering ideas;

- *Major corporations, state-run research institutions or universities* that are well-integrated into the ecosystem;

- The *user base* and its willingness and ability to try something new;

- Opportunities for the R&D community to engage in the *open exchange of ideas*, i.e. the use of 'clusters' to encourage and promote mutual inspiration;

- Last but not least, the *role of government programs in promoting and protecting innovation and intellectual property* must also be mentioned.

This chapter will aim to shed more light on China's potential as a digital frontrunner – especially in terms of the global role that China is capable of playing in this area.

9.2 Market Size and Consumer Usage Potential

That China is a global leader in consumer mobile usage has been common knowledge for a while now: Not only are the cell phones in active use (1.2 billion) almost as numerous as the population (1.3 billion), but China was also the first country in

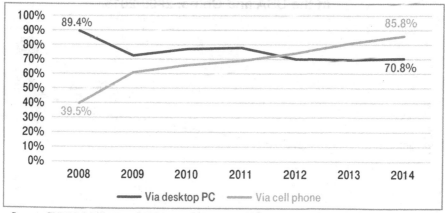

Percentage of all Internet Users in China

Source: CNNIC Survey Report 2009–2015

Fig. 9.1 China – Mobile Internet access vs. PC Internet access

the world to announce (in 2012) that more people use their cell phone than their PC for Internet access (cf. ◉ Fig. 9.1). In 2014, over 400 million cell phones were sold in China – a figure that also demonstrates the willingness of end users to purchase state-of-the-art technology and a personal stake in the very latest developments. That this has more to do with acquiring status symbols than an infatuation with technology is of secondary importance. The end result is that over a third of cellular telephony users bought a new mobile device in 2014.

Thanks to China's protectionist strategy, its market for the mobile Internet is also well-defended: In both the network provider segment and the cloud services sector, foreign companies are forbidden from positioning themselves in the market as independent providers. Only a handful of local providers are used for chat, blog and social media services, with US corporations such as Google and Facebook being effectively excluded from the market by the "Great Firewall of China." This state of affairs has not only promoted large Chinese companies such as Sina-Weibo, Alibaba and Tencent to market leadership but has allowed them to offer consumers a new and more diverse range of services, rather than simply copying Facebook, Twitter & Co.

The huge growth rates enjoyed by Internet companies speak for themselves, and since 2014, when Alibaba successfully completed what was to become the largest IPO in history, an awareness of this status quo has also been growing in the Western world. Both Alibaba and Tencent have seen revenues surge in the past, and their growth continues unabated to this day. China has been the world's largest e-commerce market for some time now.

The primary beneficiaries of the de facto walling-off of the Chinese market for Internet and cloud services have been the Chinese companies in these sectors. Insulated from foreign competition, solutions and innovations have developed that are unquestionably global leaders in their fields.

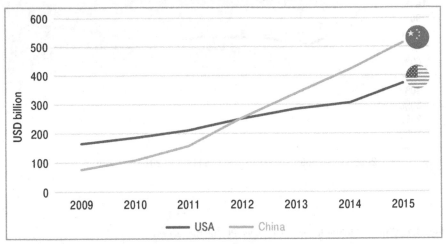

Value of E-Commerce Transactions
in the USA and China, 2009-2015

Source: KPMG Analysis of US and Chinese E-Commerce Data from Statista, Bain & Company

Fig. 9.2 E-Commerce Transaction US and China

9.3   B2B

In 2015, the B2B market in China grew to encompass the almost unimaginable size of 1.3 billion US dollars with three quarters of total e-commerce revenue (i.e. goods that are either sold or distributed via e-commerce) being earned in this sector (cf. ⊚ Fig. 9.2).

Unlike Europe, therefore, China has managed to establish its own search engines, social media services and leading e-commerce providers – with more than a little help from state protectionism. The sustainability of this success will be decided not only in terms of global competition (e.g. WeChat vs. Facebook and Instagram) but also in the continuous enhancement of these products, including improved access for the global developer community. Ideally, a frontrunner is both a market and a technology leader – and this is something China still needs to demonstrate in a global context.

9.4   Industrial Trends in Digitalization

9.4.1   Industrial Use of Information Technology: A Baseline

In terms of the industrial use of information technology, China has yet to take the lead. Although many companies deploy ERP (Enterprise Resource Planning) systems and a high degree of automation, a look at the annual reports of the major multinational software and hardware system suppliers tells us that the markets in Europe and the USA are still dominant for this sector. And yet: China has already muscled its way into the Top Five at almost every global IT corporation and it is surely only a question of time before China also becomes the largest buyer for this sector.

As noted in the introduction, however, while consumption and size may bring substantial benefits of scale they do not necessarily position China well for the role of trendsetter. Deloitte's 2013 Global Manufacturing Competitiveness Index Report awards indexes of 9.47, 8.94 and 8.14 to Germany, the USA and Japan, respectively, while China trails behind with an index of 5.89. For too long, "cheap China" has been content to leverage production scale and labor cost advantages, and rely on technology transfer. This has not only resulted in the gradual erosion of competitiveness but has meant insufficient attention has been given to the necessity for continuous, broad-spectrum innovation within production.

A currency that has steadily strengthened from year to year combined with significant increases in wages is now working to place China at a disadvantage vis-à-vis many other Asian countries in terms of unit labor costs. The (to an extent politically motivated) trend in the USA for bringing production "back home" has also served to strongly focus the minds of corporate decision-makers on China's shortcomings here. Apple's 2012 decision to relocate production of its Mac series computers back to the US from China made headlines the world over.

This was naturally recognized by China, which used politics and PR in an attempt to play down the incident.

9.4.2   Chinese Government Programs for Achieving Digital Market Dominance

China has recognized the importance of digitalization both for its industry and its people, and has adopted various strategies aimed at achieving both domestic and international predominance.

China's non-stop focus on driving mobile network expansion over the last few years means that it now not only has the most cellular network users in the world – and that by a wide margin – but is also the first country in which cell phone-based Internet access now exceeds PC Internet usage. In addition, the Chinese government has actively encouraged its citizens to make greater use of the Internet and offers many state services via apps.

This has led to fundamental changes affecting social life: In some cities, providers and consumers have become so used to phone-based services that it has now become very difficult to hail a cab without the right taxi app. And there are probably very few Chinese of the younger generation who haven't yet shopped online on Taobao or T-Mall – the world's biggest marketplace for almost anything produced worldwide. According to figures from A.T. Kearney, online shopping (B2C) volume totaled about 400 billion US dollars in 2014 and is forecast to rise by 25 percent annually until 2017. Most customers in the B2C segment are aged between 18 and 35: Overall, some 300 million Chinese make online purchases on a regular basis. This makes China the world leader in online retail, with a volume already twice the size of the US market.

Strong support is provided by the government in the form of the MIIT (Ministry for Industry and Information Technology), while the Ministry for Cybersecurity simultaneously keeps a watchful eye on anything (and everything) else in the online world.

Nonetheless, not only were many Chinese companies early pioneers in the field of mobile payments (WeChat Payment Wallet, Alipay), but many business processes were in fact digitalized for the end customer earlier in China than in the USA or Germany. The government in Beijing is therefore attempting to balance massively increased use of the Internet – and social networks and e-commerce in particular – with equally pervasive control of its content.

While social media is used by government agencies and party officials as an effective means of communicating with the population, it is also a useful tool for learning about trending topics (e.g. dissatisfaction about local grievances) early enough to be able to counter them appropriately.

At the same time, there is also an official Internet censor tasked with keeping the network free of "harmful influences" – which is naturally a very broad-based job. Everything posted online as "content" in China must either be officially approved or at least comply with a set of clearly-defined parameters. These rules, which are published openly and therefore common knowledge, can work to delay content processes – especially as regards the exchange of information or the dissemination of knowledge.

In terms of industrial digitalization, the key focus since 2014 has been on the two major government programs of *Made in China 2025* and *Internet+*.

Internet+ is expressly dedicated to the principle that the Internet (and IT as a whole) must develop beyond the conventional idea of hardware and software towards a triple concord of hardware, software and data. This involves the systematic integration of cloud computing, big data, the Internet of Things and mobile Internet. This might sound run-of-the-mill at first – as "mobile," "big data" and "cloud" can now be found (in various constellations) in almost any company presentation.

However, China's integrated approach of attempting to harvest the various trends or counter-trends and simultaneously harness their utility for business (the finance sector is a key target here) is unquestionably superior to that found in many other countries. This is then implemented in a series of practical programs – such as the financing of 100 pilot projects. Or the new free trade zone in Shanghai, which is intended to exploit new opportunities for cloud computing without the restrictions that apply in the rest of China.

The declared objective is to achieve the position of a global leader in the use of IT and in digitalization – in both industry and day-to-day life – within the next five years. This also chimes well with Gartner's forecast, which expects the size of the total public cloud market to reach 20.7 billion US dollars by 2018, assuming an initial size of 5.3 billion US dollars in 2013 and a compound annual growth rate (CAGR) of 31.5 percent (cf. Gartner 2014).

Planned and published outside the typical five year period, *Made in China 2025* makes a highly unexpected break with tradition, a fact that highlights the importance of this initiative for the government. Essentially, the program involves "upgrading" Chinese industry to handle the new era. The initial scope of the

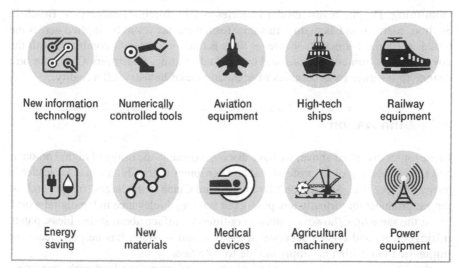

Source: People's Daily Online

Fig. 9.3 Ten Key Sectors "Made in China 2025"

program is fairly broad, adopting a general approach to improving the quality of Chinese products and increasing the efficiency of production processes. Since many companies (due to the long period of very low labor costs) are only beginning to consider automation (and therefore the transition from Industry 2.0 to Industry 3.0), digitalization has a significant but not all-encompassing role to play in many scenarios (cf. ◉ Fig. 9.3). The program's core principles are "quality, not quantity," ecologically-viable improvements in manufacturing, the optimization of manufacturing structures and talent development.

Attention is not focused solely on efficient production control or the introduction of "cyber physical systems," as they are referred to in the Industry 4.0 whitepaper from Acatech. Another key goal is to improve the integration of Chinese production in global value chains and thus, naturally, to increase the proportion of the value thereby created. In many factories operated by FoxConn, the Chinese iPhone manufacturer, this proportion is 1.9 percent: The lion's share of value creation (and also the profit) thus remains in the hands of other companies – and first and foremost Apple.

Rich rewards lie in store for those Chinese producers who manage to improve the integration of domestic production with their buyers (and end customers, as appropriate).

Here, the holistic approach taken by *Made in China 2025* will certainly help to lay the foundations for a reform of the Chinese manufacturing industry.

In summary, although China's shortcomings in terms of competitiveness across global markets are clear to see, the government in Beijing is nonetheless making huge efforts to modernize Chinese manufacturing. This has already resulted in clear improvements to productivity and should help ensure the same for quality in the next phase. The degree to which funds are in fact allocated and whether manufacturing successfully weathers the reforms are aspects that require careful monitoring. Despite a few model factories run by foreign producers – Bosch in Suzhou, Siemens in Chengdu – this transformation has yet to be applied across the board in China. Ultimately, however, it is economic pressure, combined with the government's firm convictions and earmarked funding, that offers China the best chance to leverage its economies of scale and make Industry 4.0 a reality.

9.5   Innovation

In the topics discussed above, we have discovered that in the usage of digital products and the networking of its professional and consumer users China is a clear leader in terms of sheer size. And it won't be long before China is the largest market and the largest producer for virtually any product or service of relevance in the digital world.

At the same time, however, stories continue to surface about stolen ideas, patent infringements and copies accurate in every detail – which has helped create the impression that China is simply unable to innovate.

Yet China certainly has an excellent chance of taking the lead with innovative products, and especially in the process of digitalizing its economy, for the following reasons:

1. **Market dominance will force increasing numbers of companies to disclose knowledge** (technology transfer) and Chinese companies will catch up rapidly. It's no secret: Anyone wanting to enjoy long-term success in China must be open to technology transfer. The Chinese government makes no bones about its declared intention of allowing market entry to strategic industries (such as aircraft construction, rail, IT) only on the condition that the corresponding products are also manufactured in China or that at least part of the value chain is located in China. In the end, most companies agree to these terms since China is a market that is simply too important to be ignored. Things won't be any different for digitalization: In this context, source code may need to be disclosed (due to security concerns, for example) or certain services (especially within networking and cloud computing) will be marketable only in tandem with a Chinese JV partner.

   Ultimately, a lot of knowledge will change hands. And even if companies manage to design new products in ever shorter cycles, the Chinese partner will of course be trained as well, Chinese employees will gain new knowledge and colleges, too, will include the content in their programs. In this way, the foundations for local technology development can be laid (cf. railway construction or solar technology).

2. **The protection of patents and intellectual property rights** is taken far more seriously today in China than in the last decade. The Chinese government has long since realized that the defense of intellectual property is in its own interest. For many years after the opening up of the PRC, protecting intellectual property was a low-priority goal. After all, it contradicted the core principle of socialism, namely that "property" only exists in a very restricted sense. Why then, should the ownership of abstract ideas be protected?

Ultimately, however, repeated pressure from overseas partners and then Chinese companies themselves eventually led to a rethinking of policy here. China realized that patents are outstandingly amenable to commercialization, and that a lack of effective protection for its own products in global markets and within its own borders stood in the way of its globalization goals. Intellectual property consultants Wurzer & Kollegen provide a neat summary of the situation and managing director Alexander Wurzer himself sees the developments as clearly heralding China's entry into an entirely new phase in terms of handling intellectual property: "China's relationship to intellectual property in phase 1 was initially characterized by extensive counterfeiting and copying on the part of Chinese companies. In phase 2, this transitioned into the steady buying-in of foreign technologies and licenses, which increasingly enabled the former counterfeiters and imitators to position themselves as legal and innovative competitors. The report documents China's entry into a new, third phase: the systematic establishment of an international patent portfolio based on domestic innovations that is suitable for targeted deployment against competitors by exercising rights of prohibition, actions for injunctions, or by demanding the payment of compensation or license fees."

Chinese companies (and in the field of information technology, these companies are legion) are highly competent at protecting themselves and successfully deploying their own patents. As a result, both domestic R&D and domestic innovation have now been accorded an entirely different status. This is also highlighted by the rising number of patent applications (patent registrations are now significantly higher in China than in the US; cf. ◉ Fig. 9.4). In July 2015, *Wirtschaftswoche* magazine reported that the 2500 patents registered since 2013 by China related to enabling technologies for the Internet of Things (IoT) amounted to double the US patent volume and six times that of Germany. One should note that the level of innovation represented by such patents is not always comparable, however. One final word on patent protection in China: Recent instances of patent litigation, such as the cases brought by German SME Herting in January 2015 (alleging infringement of patents for specialized electrical connectors) and Apple in 2014 (related to 'Siri' voice recognition) have been decided in favor of the overseas plaintiff. This, too, signals the general change in attitude and further strengthens the trend towards greater innovation and legal protection.

3. **Increased spending on research and development**. After the USA, China is already the country with the second-highest level of spending on R&D. Should China achieve its goal of ploughing at least three percent of GNP back into research by 2020, then, according to figures from the OECD, it will be the country with the highest level of R&D spending (cf. see ◉ Fig. 9.5).

For many years, Beijing's focus was on the creation of infrastructure and production capacity. Yet this strategy is now no longer an adequate guarantor of healthy economic growth into the future. Accordingly, a rethink has been taking place at all levels of government, and funding in R&D has recently been significantly increased (approx. 300 billion US dollars in 2014 alone).

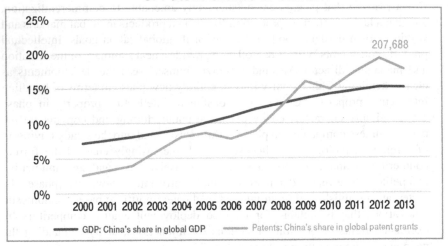

Source: Patents Share

Fig. 9.4 China's Innovative Capacity

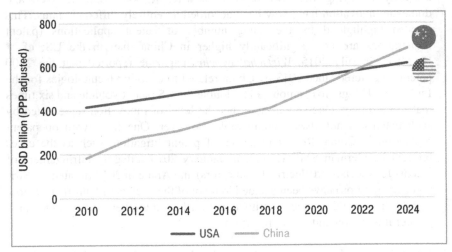

Source: 2014 R&D Global Funding Forecast from Battelle

Fig. 9.5 China vs. USA: Outlook on Research & Development Spending

Since 2014, there has also been a particular focus on the digitalization of manufacturing. One should expect that China will make every effort possible to equip its higher education and research institutions with a level of funding that is appropriate for attracting international experts and upgrading technical facilities to the technological state of the art.

4. **Systematic improvements to durability, simplification and processes.** Why has China never created an iPhone? Why was the PC developed in the USA? Why was the MP3 format invented in Germany? While often asked, these questions all suffer from an over-simplification of the larger picture: Only a handful of innovations are actually cited (and in particular those that have been successful on the consumer market) and the timespan considered is very short (the last 20 years). If we were to expand the time frame to, say, 3,000 years, then China instead becomes the land of continuous, game-changing innovation. Even if we restrict our view to the last few decades, however, while we might not be able to identify a product that everyone in the world now uses, it still remains a fact that the phones or tablets we enjoy (and buy for no more than a couple of hundred euros) are the result of Chinese factories using highly innovative manufacturing processes. And these processes, tools and quality assurance systems are protected by a broad portfolio of patents. In many cases, products are also developed further to meet the needs of the Chinese market – the aim often being to achieve a less complex and/or more rugged product design while not compromising on performance. Chinese high-speed trains are an excellent example in this context. While a good deal of German technology comes along for the ride (e.g. from Siemens or brake manufacturer Duerr), a considerable amount of money (and R&D effort) has been spent on making locomotives more reliable and reducing their maintenance needs, etc. "Good enough, but at a much lower price point" is the maxim, and this has been China's secret weapon in global markets stretching from Turkey to Brazil and across the African continent. In May 2015, the *Handelsblatt* business newspaper reported that Deutsche Bahn was also considering China as a supplier of trains and spare parts. If we consider the digital factory of the future, China stands to benefit hugely from this strategy. After all, future technology must also be manageable by human users: If it can avoid the pitfall of hopelessly overburdening its operators, then it will also be less error-prone. The digital factory of the future will not necessarily need the most expensive plant networked with a full array of cutting-edge technologies, but will deploy an intelligent system of proven components while still approaching the limits of what is technically possible.

5. **Beijing's clear focus and market player bundling.** The highly relevant government programs *Internet+* and *Made in China 2025* have been discussed above. Complementing these, China has also established a series of highly-focused programs aimed at driving the digitalization of the economy. While the effectiveness of such programs can be debated and – as with any government program – misallocation of resources will no doubt be rife, these programs work to create one thing above all else: focus. This gives all of China's political and economic stakeholders the shared objectives of making better commercial use of the Internet while simultaneously modernizing industrial manufacturing. These are, after all, two complementary goals that offer many opportunities for working together. Chinese companies will (with gentle encouragement from "above") utilize these opportunities and – especially when backed by the government – attempt to bring a large number of international partners on board. This can be

arranged in the form of collaboration agreements with a few partners. Then again, some Chinese IT companies have also joined the major global industrial associations, as is the case with Huawei, which is now a member of the Industrial Internet Alliance founded by US companies. This will lead to further knowledge transfer – whether involving standards, security models or many other fundamental topics. From this starting-point, China will develop its own version of the networked Industry 4.0 (as it has of course already done very successfully in B2C) and, thanks to the great agility of its companies, will aim to "upgrade" these new technologies into global standards by means of exports. This future can already be seen in current developments. Customers can use the Chinese "Unionpay" card in any shop in Düsseldorf or Frankfurt. Many German online shops also accept "Alipay" payments. We can therefore assume that the same will happen in the field of networked robots, plant systems and customers. Perhaps not as a unified standard (as the US would certainly move to block such an outcome) but as an alternative standard.

6. **Talent (local and international).** Sustainable progress depends on having the right subject specialists, experts and visionaries (in moderation). Silicon Valley is the textbook example of how a mixture of first-class universities, capital, and domestic and international migration has created a one-of-a-kind high-tech ecosystem. While China is investing heavily in higher education, it is struggling to cover its needs for subject specialists. Outside the Top 20 universities, education is also significantly less comprehensive and state-of-the-art. A further problem is that too much knowledge is simply crammed and learned by rote and too little emphasis is placed on creative thinking. Then again, China is benefiting from the increasing appeal of its labor market – especially in its high-tech clusters. This is bringing a lot of Chinese top talent back to China. With considerable improvements being made to living conditions, foreign experts are now also flocking to China's larger cities. And local salaries for top executives are now significantly higher than those paid in Germany.

Yet for a radical process of digitalization it is not so much software developers and IT experts that are lacking but, ultimately, the skilled workers who can bring all these plans to fruition. A prodigious gap is opening up here, and it probably cannot be closed quickly enough to keep pace with rising demand. Compounding this problem, we can also see how shortcomings in the exchange of ideas, concepts and new solutions on international platforms or blogs are working to isolate users and software developers in China from important sources of knowledge. This will work to disadvantage China in the long term, since its "Great Firewall" (like the erstwhile Great Wall of China) will act as a barrier to the rest of the world, and neither students nor experts will be fully aware of events beyond its borders. This is perhaps one of the greatest obstacles to China's plans to become a leader in the digitalization of manufacturing and business in general: Openness and the unimpeded exchange of new ideas about interesting technologies, shared, cross-border development on common software projects – all of this will be needed if China wants to prove its mettle as a global digital frontrunner and not as a clever "nerd" stuck behind a wall.

9.6   Conclusion

Is China really a digital frontrunner? Applying the criteria listed in the introduction, we can draw the following conclusions.

There can be few doubts about the tech-savviness of Chinese customers – nor of their numbers: In key Internet-related sectors (B2B, e-commerce, etc.), China is currently the world's largest market. We can also assume that the same will very soon be true for the entire spectrum of products for the 'Internet of Things' (IoT). Starting from this strong position, there will be a considerable need for technology and know-how, some of which will be satisfied by transfer from foreign technology leaders and some (increasingly) by development within China itself. More and more Chinese customers will also be expecting – and making use of – innovative enhancements.

China is also rapidly upgrading its ecosystem: Universities, companies and state-run R&D institutions are increasingly pooling their resources and benefiting from such cooperation. Startups are also becoming much more numerous – some with international co-founders who act as stimulators in their sectors. The ecosystem now has everything it needs.

There has also been a real paradigm shift in government programs and the funding of innovation. If resources are properly deployed and intellectual property rights are concurrently strengthened, this is an area with a lot of potential. Initial efforts here are on the right track.

The innovative strength of the economy at large is a more problematic issue, however. While China does have its innovative hot spots, such as Beijing, Shanghai/Nanjing, Xi'an or Guangzhou/Shenzhen, true clusters cannot really be identified – especially considering the size of these economic areas. While the sheer volume of patents and businesses is impressive, calculation of the figures per student, facility or company shows that a comparatively small country such as Israel is far more innovative. In the wider economy, SMEs in particular are still lagging very far behind.

Turning finally to consider community, talent and the sharing of ideas, this is where the key challenges for digitalization are to be found. China is unlikely to be able to master the enormous challenges involved in educating and training the skilled professionals and workers required for a comprehensive reform of the manufacturing sector. Equally, the firewalling of the entire Internet in today's day and age (in which students source a large part of their current knowledge from the web) is very counter-productive. Not least because the filters that "protect" the Chinese Web are fairly heavy-handed and block a lot of useful content. Speed losses are also enormous, making it virtually impossible to engage in rapid data exchange with the global business community.

This is where China really has its work cut out, and it will be interesting to see the home-grown solutions developed here.

Industry 4.0 is the great opportunity for Chinese manufacturing to make a quantum leap in terms of its modernization and leapfrog over an entire industrial era. If this proves successful (bolstered by government programs, massive spending, improved education and the migration of highly skilled experts), then China will mature into the global market and innovation leader in digitalization.

As for foreign companies, these should lose no time in getting involved and participating in the trends that are already being set within China today. Only those who engage (and partner) with the Chinese competition on its home turf today will know their competitors in the global markets of the future.

References

2014 R&D Global Funding Forecast from Battelle. Accessed October 5, 2015, from http://battelle. org/media/global-r-d-funding-forecast

Acatech (2013, April). Umsetzungsempfehlungen für das Zukunftsprojekt Industrie 4.0.

Acatech (2014, March). Smart Service Welt.

Alibaba's Prospectus. Accessed October 5, 2015, from http://hsprod.investis.com/shared/v2/ irwizard/sec_index_global.jsp?ipage=10351468&ir_epic_id=alibaba

Chan, J., Pun, N., & Selden, M. (2013). The politics of global production: Apple, Foxconn and China's new working class. *New Technology, Work and Employment, 28*(2), 104–105.

Gartner (2014). Report *"Emerging market analysis: China, nexus of forces trends and opportunities"*.

KPMG (2014). *E-commerce in China: Driving a new consumer culture.* Accessed September 24, 2015, from http://www.kpmg.com/CN/en/IssuesAndInsights/ArticlesPublications/ Newsletters/China-360/Documents/China-360-Issue15-201401-E-commerce-in-China.pdf

KPMG Analysis of US and Chinese E-Commerce Data from Statista, Bain & Company. Accessed October 5, 2015, from http://www.kpmg.com/CN/en/IssuesAndInsights/ArticlesPublications/ Newsletters/China-360/Documents/China-360-Issue15-201401-E-commerce-in-China.pdf

Patents Share: World Intellectual Property Organization, IP Statistics Data Center, March 2015; GDP Share: International Monetary Fund, World Economic Outlook Database, 2014.

Ten Industries Prioritized for Upgrading Manufacturing Ability. Accessed October 5, 2015, from http://en.people.cn/n/2015/0930/c90000-8957295.html

Wurzer, A. (2013). *Drei Phasen: China auf dem Weg zur Innovationsschmiede.* http://www. bvmw.de/landesverband-hessen/kreisgeschaeftsstellen/darmstadt-darmstadt-dieburg/news-detailseite/artikel/chinas-patentstrategie-bedroht-deutschen-mittelstand.html?L=0

Clas Neumann is Senior Vice President, Head of Global SAP Labs Network and Head of Fast Growth Market Strategy Group at SAP. As Head of Fast Growth Markets since 2013, he has been orchestrating both capital spending and Group-wide strategy for the development of SAP's business in regions offering potential for rapid growth. As a member of the SAP Senior Executive Team, Neumann can look back on a 20-year career at SAP. 14 years of this period were spent in India and China, where he decisively influenced SAP's entry into the Chinese market. He has also directed the Global SAP Labs Network since 2009. Prior to this, he was President of SAP Labs India and as SVP was responsible for software development, managing technical teams based in China, Germany and India. He managed the establishment of SAP's R&D Center in Bangalore and its further expansion to the present team of 4,000 engineers. Neumann is a spokesperson for the Asia Pacific Association of German Industries, a board member of the East Asian Association and a member of the German-Indian Advisory Committee to the Indian Prime Minister and the German Chancellor. His is the co-author of several books. Neumann lives with his wife and three children in Shanghai, China.

The Cloud Drives Harmonization and Standardization

10

Klaus Hardy Mühleck

IT islands – they rise up from the flood of data in every corporation and every medium-sized company. As sunny as they may sound, they cast a long shadow over a company's flexibility. In this dynamic age of digital transformation, effective administration is wrecked on the shores of these islands. In reality there are often local databases, each of which has grown independently for years and now holds layer upon layer of a quarter century of IT history. These IT landscapes are almost impossible to manage efficiently, and their complexity leads to inefficient and opaque processes which waste the valuable energy of IT employees and CEOs alike. It starts in day-to-day business: Existing data is needlessly recorded multiple times, and relevant information is not taken into account in new developments because no one knows it exists – or where it can be found. Ergo, even as companies merge and become one, their IT remains fragmented across different sites. This was the case at thyssenkrupp – and we wanted to change it. International companies in particular require agility, flexibility, global communication and cooperation in order to gain a competitive advantage. One of thyssenkrupp's goals, therefore, is to consolidate and harmonize its existing IT landscape worldwide.

10.1 Flexible IT with the Cloud

To break open rigid structures, you need a consolidated enterprise resource planning (ERP) backbone as well as a comprehensive solution: the cloud. The cloud creates a shared platform and offers an opportunity to ease the strain on your in-house IT. Furthermore, the right IT partner can ensure global availability and increased flexibility. This allows the company to focus on its core areas of business: engineering, production and logistics. In global corporations such as thyssenkrupp,

K.H. Mühleck (✉)
thyssenkrupp AG, ThyssenKrupp Allee 1, 45143 Essen, Germany
e-mail: klaus.muehleck@thyssenkrupp.com

© Springer International Publishing Switzerland 2017   105
F. Abolhassan (ed.), *The Drivers of Digital Transformation*, Management for Professionals, DOI 10.1007/978-3-319-31824-0_10

this business is handled by different and sometimes alternating sites – wherever there is available capacity, resources and customers. But the data always has to be right where it's needed, and it has to be available in formats that can be further processed. Resource planning and customer management cannot be limited to specific sites or departments because this impedes business. Agility and flexibility, by contrast, can be achieved through standard applications which can be used in the cloud in the form of software-as-a-service (SaaS) and can directly access the available data, as well as through workplaces that are no longer tied to local servers and user accounts. With centralized administration, it takes no extra effort to ensure that all employees are working with the same version of applications, all necessary updates are installed regularly, the data remains compatible and no security gaps are ignored or go undetected locally.

The fourth industrial revolution, which follows three previous revolutions – mechanization, mass production and digitalization – relies on the factors of information and communication. Whether the "Internet of Things" means that things are connected to each other or humans are included in this exchange, the smooth flow of data is what will guarantee business success in the future. This is both a challenge and an opportunity for the economy. And it is such important terrain that competent, integrative partners are called for. This is why thyssenkrupp opted for T-Systems. The planned global harmonization of thyssenkrupp's data and processes and the standardization of its workplaces required the kind of high availability that T-Systems can offer with its twin-core data centers. It was necessary to find a stable solution which guaranteed the reliable use of standard components and the migration of existing company solutions to the cloud, thus doing justice to the diversity of the company and its multinational business areas.

10.2　The Starting Point

thyssenkrupp is no longer associated solely with steel and iron. Steel Europe and Steel Americas are just two of the group's six business areas. The rest is handled by the other business areas: Components Technology, Elevator Technology, Industrial Solutions and Materials Services. The group consists of around 500 companies with sales of about 41 billion euros, which means that it's a heavyweight in the industry. And so is the project that thyssenkrupp launched at the start of 2015 with T-Systems: A good 80,000 computer workplaces and around 700 data centers and server rooms will be moving to the cloud in the space of 36 months. The project to turn thyssenkrupp into an integrated corporation involves more than 155,000 employees at about 1,700 sites in roughly 80 countries.

10.3　From the Public Cloud to the Private Cloud

thyssenkrupp aims to establish a group-wide integrated IT landscape which makes global communication and cooperation easier and more efficient while also meeting the different IT requirements of each business area. The cloud is the ideal vehicle for this. The easiest way to get started is by using the public cloud, which offers the chance to launch and test applications. The public cloud provides access to webmail and virtual drives as well as applications, collaboration services, platforms and infrastructure services. In Asia, for example, thyssenkrupp uses Microsoft Office 365 to keep more than 10,000 mailboxes in the public cloud. But the cloud can do much more: In an IoT lighthouse project, started in the USA, Germany and Spain as pilot countries, thyssenkrupp Elevator has been using a cloud for the Industry 4.0-connectivity of its elevators. By means of predictive analytics, the data collected from the elevators is analyzed so that flexible maintenance intervals can be planned in advance, which massively reduces downtimes and transportation standstills.

The public version of the cloud is just the first step into the cloud for thyssenkrupp, however. The overarching strategy behind the group's "unite" IT consolidation program is to migrate more than 500 domains and active directories, around 700 data centers and server rooms and over 10,000 applications from the decentralized IT landscape to the private cloud of T-Systems in order to ensure highly secure, standardized, global IT availability. The private cloud offers two key advantages over the public cloud here: solutions tailored to thyssenkrupp for protecting the group's intellectual property and personal data (keywords: data protection and information security), and dedicated access rights which mean that only company employees or authenticated users can access and use the IT infra-structure and its applications. These include services which are already in the public cloud, such as the maintenance service of thyssenkrupp Elevator, an important division with over 50,000 employees.

10.4   Greater Efficiency through New Architectures

The cloud is therefore expected to deliver a lot in terms of performance. But what do you do if the cloud cannot manage this? An elevator is a good example of the potential offered by the cloud: An elevator not only connects floors, it connects people. And, just like an elevator, data and applications have to be where they are needed – the faster, more directly and more securely, the better. If you are prepared to rethink architectures and standardize (data) structures, you can give a boost to efficiency and workflows. This kind of flexible construction doesn't only exist figuratively, however – in the case of the elevator, we can see it in the "Multi" concept. Multi is a brand-new elevator technology developed by thyssenkrupp which will bring greater flexibility to the construction of high-rises in the future. Skyscrapers have always been constrained by elevators being pulled vertically through their cores, but the new Multi elevator concept fundamentally changes this. With Multi, the elevator comes to the passengers and picks them up where they are. It is based on a technology that does away with the familiar rope and instead uses magnetic levitation to send cabins through a building. This makes it possible for elevators to move horizontally as well. Not only can the elevator navigate flexibly, it can also act as part of a group along with other elevators, which can dodge one another and be requested at specific times – peak times, for example, such as the start of the working day or the arrival of a bus for an event. This is Industry 4.0 *par excellence*: The application eliminates crowds in front of elevator doors by evaluating traffic flow data, while preventive maintenance reduces downtimes. And all of this is made possible by the cloud.

Above all, Multi illustrates a key point: Technologies are entering our everyday lives in a variety of ways. But they will only add value if they are available and functional. The cloud lays the foundation for this. It also lays the foundation for more flexibility: Just as Multi allows for new architectures that have never existed before, the cloud allows for network architectures that were previously unthinkable. And – just like the elevator – it ensures more flexibility, availability, speed and safety.

10.5   Dynamic Workplaces from the Cloud

Flexibility is a decisive advantage of a company-wide cloud solution: Data is accessible, applications can be used, and documents and information are available whenever an employee needs them. This is what thyssenkrupp has achieved using the Dynamic Workplace from T-Systems. This solution gives employees permanent, secure access to their personal workplace in the T-Systems cloud. Workplace operating systems and desktop applications are virtualized, so the user is completely independent of a specific end device or mobile operating system. In order to access the system, users must authenticate themselves, which they can easily do in a browser or an app. Users operate their workstations in much the same way as before – something that was important to thyssenkrupp. This means employees do not need to familiarize themselves with an entirely new digital working world, new workplace environments or programs. Instead, they can use familiar applications that have simply migrated to the cloud while at the same time enjoying the benefits of a new infrastructure that offers permanent access to their stored data. On top of this, there is more security – regardless of the end device.

The variability of the end devices leads to increased mobility for users as well as simple administration, because new workplaces can be created easily. All workplaces are virtualized centrally in T-Systems data centers, where all of the applications run and all of the data is stored. This significantly reduces the cost of hardware for thyssenkrupp, particularly for maintenance, system set-up, software installation and updates. Because the processing takes place in the cloud, no additional demands are placed on the hardware.

An approach like this paves the way for cost-efficient end devices, but it also offers the option of continuing to use existing PCs. This is a particularly important factor in the transition phase, which can last for years in a big project like this. It means that employees can keep using the old, decentralized system environment while the new infrastructure is being established and the applications are gradually added. As a result, the IT infrastructure can be harmonized and standardized without interrupting functioning business processes.

10.6 The Location Question

It was clear relatively early on that thyssenkrupp did not want to run its own data centers in the long term. The complexity of this – which would continue to grow over the years – would simply tie up too much capacity and money. This is where the company's IT partner comes in – a partner who not only has the necessary expertise but also has the latest technologies and enough data center capacity. This spares valuable capacity and financial resources which can be applied to the group's core business instead.

As an international company, it was essential to thyssenkrupp that its IT partner have data centers not only in Germany but also internationally.

For its unite consolidation program, thyssenkrupp has access to five twin-core data centers run by T-Systems. The data centers for Europe are in Frankfurt, the ones for the USA are in Houston, the ones for China are in Shanghai, and there are also data centers in Singapore and São Paulo. This is another point in favor of the private cloud: You can specify where certain data will be stored and where special applications will run. This kind of controlled local data storage makes it possible for a global company to comply with the laws and official requirements of each country. For companies operating worldwide, this issue relates to the protection and security not only of the company's own sensitive data, but also of the data associated with local partner companies or individuals – employees and customers alike. Data protection needs to be guaranteed in accordance with national laws and guidelines.

Since it may be mandatory to disclose stored information depending on the location, it is critical to meticulously plan who may access which data from where and in what way. The question of where this data is physically stored or – even more importantly – where it absolutely must not be stored determines whether company secrets remain secret or not. This also applies to cases such as China, where official authorization is required before an approved encryption technology can be used. With T-Systems data centers in both the USA and China, thyssenkrupp is able to separate its data in order to ensure that it is protected in accordance with the requirements of each country. This "grounded" data cloud makes it clear that the term "cloud" is sometimes too imprecise. Information is not actually processed and stored in a cloud in some indeterminate location. Instead, a company must be able to precisely pinpoint the location of its entire inventory of data and the applications that access it. Thanks to the T-Systems data centers – which take account of local requirements and specifics – data can be processed collaboratively in many different countries, from the USA to Europe and Asia.

10.7 Top of the Agenda: Data Security

thyssenkrupp benefits from the ability to trace and control the international flow of data, from the strict data protection legislation and EU regulations that govern German data centers and from the stability and high availability of the twin-core data centers. The modern technology behind these fully redundant twin data centers guarantees the greatest possible availability and security. If a disruption unexpectedly occurs in one of the data centers, its twin core immediately takes over operations. This should be a top priority for a global corporation, because a company-wide workplace solution such as the Dynamic Workplace is only reasonable and economical if the availability of the applications and data in the data center is assured.

The security aspect also relates to protection against external attacks. After all, the German domestic intelligence service estimates that the annual damages from electronic espionage in Germany amount to around 50 billion euros. To protect its network as much as possible, thyssenkrupp recently adjusted its infrastructure, defined a four-level security model and invested in protecting its clients and zones. This security management program is being further refined as "Secure Cell Management 2.0" together with T-Systems in accordance with German law in order to safeguard the company's data. The administration of users and profiles remains in the hands of thyssenkrupp, while tool management and the interplay of the data centers and network components are being handled together with T-Systems. This involves next-gen firewalls as well as state-of-the-art security appliances, semantically linked CM systems and surveillance systems which monitor traffic.

The virtual concept behind the Dynamic Workplace offers increased security as well, since there are practically no local weak points anymore. Thanks to virtualization, which starts on the network level, users are strictly separated and better protected from one another and from the server. This makes it possible to centrally manage user access regardless of the device. It also improves the exchange of data between Dynamic Workplace users, which takes place in the protected space of the data centers.

10.8 Challenges as a Group

Right from the start, complexity is the biggest challenge when it comes to planning and thinking about every aspect of a project like this. Even though the cloud will eventually simplify the IT landscape at thyssenkrupp, the first task is to successfully transform the existing IT islands. In this case, the entire working environment was still distributed among various providers or at thyssenkrupp itself, in around 700 data centers and small server rooms. The Dynamic Workplace gives all employees standardized access to this environment. The obstacles lie in the transformation itself: Thousands of applications distributed in the field have to be consolidated and standardized. To do this as efficiently and stringently as possible in each country, thyssenkrupp initiated its own program (Data and Process Harmonization, or "daproh") which is running parallel to unite and is mandatory for all business areas. The aim is to harmonize processes and applications on the local level alongside the group-wide IT infrastructure consolidation. The only way to ultimately achieve a positive result is to harmonize all data and IT processes, without exception – from simple communication tools to the business data for ERP analyses.

thyssenkrupp consists of around 500 companies. As different as these companies' products and production chains are, they are all a part of thyssenkrupp. The task here is to bring together all sub-sections of the group, break down barriers using data technology and generate the greatest user value. Connections within the company's own value chains are urgently needed in order to do business more effectively, quickly and profitably as a group. Take transportation as an example: Some parts of the group handle transportation independently, while others use an overarching transportation management system. The question to be answered by the management and solved by IT is this: In the future, will each company handle its own logistics, or will there be worldwide synergies which take transportation management into an entirely new dimension? The harmonization and standardization of the processes and data in each company can be achieved most easily as a group when IT and its infrastructures are moved to the cloud.

International expertise is called for here, too: Anything associated with American or Chinese legislation, for example, will quickly bump up against country-specific obstacles, such as export control regulations. This is why it's good to have a global partner who has both a presence and experience in the respective target markets on the one hand and a feeling for the national details on the other. This opens up the opportunity for new collaborative models which the partners would not have been able to initiate as successfully on their own.

10.9 Migration in Two Steps

Even when the goal is a shared cloud, it is necessary to first take a more differentiated view of the international consolidation of group data. You cannot simply stuff everything into a cloud. This would overload the cloud and raise false expectations. Cloud computing essentially means moving data center services – i.e., server activity – to a network and providing software services in this network. Cloud computing is not a miracle, it's an organizational criterion. Cloud computing enables companies to bring software such as SAP, Oracle or homespun Unix applications into a large network instead of running them independently in the company's local data center. At thyssenkrupp, this network is orchestrated by T-Systems.

On account of the IT history of the business areas, there were many applications that were not yet ready for the cloud but needed to be. It is relatively easy to migrate from the public cloud to the private cloud with applications such as Microsoft Azure, Dynamics and Office 365, SAP HANA, SuccessFactors, Ariba and IBM WebSphere. The hard part is porting the thousands of applications across the board which are not yet in the cloud. This cannot happen overnight. The migration is therefore taking place gradually at thyssenkrupp. Gradual migration ensures that the initial plans remain realistic to the end and can be successfully realized –

because nothing is left out of the equation, all of the applications are carefully made ready for the cloud and changes can be addressed as needed.

10.10 Can It Be Done without the Cloud?

Considering this tremendous effort, you might wonder whether it wouldn't be easier to leave everything as it is and try to tackle the problems using our own networked data centers. In brief, the answer is no. This solution would have an extremely short life span and offer no future prospects for thyssenkrupp. From corporations to medium-sized companies, data only creates value when it is combined and exchanged. The better this works and the deeper the analysis, the greater the value. Industry 4.0 is more important to the economy than pithy reports would lead us to believe. It is not just about using a scancode to fill a bag of muesli according to a consumer's taste preferences, or about manufacturing a clutch on demand when a special car is ordered. It is about the ability to analyze and visualize a company's data at any time and use the results to initiate actions for optimizing business processes and customer value. All of the data that provides an insight into a company's current situation and offers a glimpse of the future through predictive analytics must be synchronized and linked. This requires a secure global infrastructure, which is the backbone of our economy. The backbones of our companies need to meet the same high standards as the goods we produce. These are very different business areas of a shared economy. It is also critical to know what you do best and where you should invest. Industrial companies such as thyssenkrupp can simply no longer afford to run a large number of their own data centers. It takes up too much capacity and costs too much money, and the market is too dynamic for an in-house IT department to fully keep up with it. Corporations and medium-sized companies alike need partners such as T-Systems for this. And these partners, in turn, are partners to Microsoft, SAP, IBM, Oracle, Salesforce and others – namely, the providers with whom industrial IT departments have always cooperated directly. This triad – industrial companies, infrastructure providers and software firms – has the potential to become a harmonious model for success.

When it comes to the cloud, thyssenkrupp has limited itself to eight to ten partners who are being orchestrated by T-Systems. It would be unrealistic to think you could manage everything with a single partner; on the other hand, it is important to ensure that individual specialized solutions and diversity do not lead to unmanageable chaos. A cloud arrangement does not require hundreds of partners. If you have eight to ten cloud platforms which cover your architecture and applications, everything can be orchestrated harmoniously.

Then you still have choices, you can coordinate processes and not be tied to a single solution. This is important because it gives you the necessary bandwidth to go down a new route. A company's IT department needs this flexibility because, ideally, it will be the technological partner to the specialist departments which will realize the necessary solutions. IT is an enabler which works best when it can almost disappear behind the solution that has been realized. The cloud comes from a provider, and the solutions are set up by the respective partners – guided by the IT department. For example, if a social business network is implemented, the head of communications and the entire community works on the issue together with the IT department and the provider. Using this approach, thyssenkrupp's IT department managed to develop a prototype of the new Enterprise Portal in just four months. The IT department was able to do the groundwork quickly and dynamically instead of trying to handle everything itself and then taking eight years to implement it. In this context, the IT department views itself as a technology partner who navigates through the architecture – just like the Multi elevator which takes flexible new routes. The role of IT is to boot up the backbone and then supply the necessary dynamics on the frontend to the partners.

10.11   Conclusion

The goal is clear: move away from traditional IT and a decentralized IT environment toward an integrated, standardized and consolidated data center solution. And do it gradually so that the new IT department, the effects of the digital revolution, the Internet of Things and the connections between people, machines and the Internet move in the direction of Industry 4.0 and generate added value for the company. I will not be satisfied until the end users say: "The result is convincing, the transition was well-supported and all of this will create added value for us and our company in the future." And I am convinced that the cloud is the right platform for us here.

Klaus Hardy Mühleck has been Chief Information Officer of thyssenkrupp AG, based in Essen, since the start of 2013. As the Director of Information Technology Management, he focuses on establishing sustainable IT structures for the diversified industrial company. thyssenkrupp consists of six business areas: Components Technology, Elevator Technology, Industrial Services, Materials Services, Steel Europe and Steel Americas. Mühleck's responsibilities include defining the company's IT strategy together with the CIOs of the individual business areas and standardizing application structures and IT infrastructures worldwide on the basis of uniform specifications. Mühleck has a degree in electrical engineering and most recently served as Chief Information Officer (CIO) of Volkswagen AG and, prior to that, as CIO of Audi AG and Mercedes Benz AG.

100-Percent Security – A Desirable Goal? 11

Michael Weppler

In the summer of 2015, we learned that German Chancellor Merkel's PC had been infected by a virus. At the same time, the entire network of the German parliament – including the computers belonging to all the members of parliament, was infected by a trojan. The initial assessment by experts was that it would take 18 months to completely remove the malware from the IT infrastructure. What can be done to prevent attacks like these? Can we not expect the highly sensitive networks operating at the heart of our democratic process to be 100 percent secure? In an era of smartphones, social media, big data and the cloud, security and confidence are issues of increasing concern to IT managers. There are many technical answers to these questions, but there are other issues of a more philosophical nature about the digital world 2.0 that cannot be solved simply by bringing in more IT, more control or more laws.

11.1 The Risk in Figures

The TÜV Rheinland technical inspection agency receives about 800,000 emails every day, but only 150,000 actually get through to the employees' mailboxes. In other words: 80 percent of all the emails received are spam, and of those that do get through, many can be deleted without a second thought. 20,000 people worldwide work for this organization on 20,000 PCs, tablets and laptops – not including smartphones. Each of these devices is therefore spammed an average of 30 times a day – although of course not every email poses a threat to the entire system.

The numbers are as impressive as they are frightening because TÜV Rheinland is a technical service provider without any large R&D departments. The company does not own thousands of patents and its assets are not in the form of blueprints for

M. Weppler (✉)
TÜV Rheinland AG, Am Grauen Stein, 51105 Cologne, Germany
e-mail: Michael.Weppler@de.tuv.com

© Springer International Publishing Switzerland 2017   115
F. Abolhassan (ed.), *The Drivers of Digital Transformation*, Management for
Professionals, DOI 10.1007/978-3-319-31824-0_11

tomorrow's high tech products: its assets are the minds of its employees, and Trojans have so far been unable to penetrate that far. Companies whose business model and competitive edge is computer-based report much higher figures.

This is why we should be more critical about, for example, the risk of interruptions to Industry 4.0 data streams. Because any interruption of the data streams to driverless cars, smart homes or smart factories in which the components control their own manufacturing process could easily result in harm or physical injury to people. These challenges have only come to light since we entered the era of exabyte (approximately one million terabytes) data storage and transmission. The cloud plays an important role here, and the question has to be asked: Can we really rely on data being available at any time, whenever it is needed?

11.2   The Inexorable Demand for Storage Capacity

Ten years ago, the cloud was something of a niche subject discussed mainly by information technology specialists. Today, very little can happen without the cloud. No matter how fast the price of terabyte memory cards falls, the demand for storage capacity grows even faster. Importantly, demand is not continuous but – like power consumption – it ebbs and flows. Power consumption is particularly high in the morning, at noon, and in the half-time break in international soccer games. A medium-sized company operating in the automotive component supplier industry may need a huge amount of memory twice a year when, for example, it is completing the development of a new component on which a global team, including the customer, is working simultaneously in realtime and in 3D. Once the component is ready, the memory is no longer needed. During this development phase, the company therefore makes use of a cloud provider and very sensibly avoids having to invest in its own data center.

The demand from the general public for storage capacity continues to rise to exceptionally high levels. For example, more minutes of film are uploaded to Youtube in two weeks than German television, including all commercial broadcasters, have produced in total since they first began. A five minute loss of service by Google would reduce the world's Internet traffic by 40 percent.

Google Maps was used by one billion users every day in September 2014. That makes the data-intensive map service the most successful product in human history.

The current contents of YouTube would probably have filled the world's entire available disk storage space 15 years ago. And there is no reason why this trend should not continue to gather pace.

11.3   Reliability Is Vital

When I bought a 386 computer in the early 1990s, my local PC dealer told me I was making a purchase for life. Five years later, the only use I could find for the 386 was as a glorified typewriter; the one MB hard drive seems almost prehistoric from today's perspective.

The positive side of the fall in prices is that space is no longer a limiting factor. Just 5 years ago, hardware was so expensive that cloud providers wishing to offer 99.99 percent availability – which was considered theoretically possible – could not have survived in business because customers were not willing to pay the price. Just to clarify: 99.99 percent is the equivalent of a maximum of 52 minutes downtime per year. In its "Zero Outage" quality assurance program, T-Systems now offers 99.999 percent cloud availability – the equivalent of a maximum downtime of about five minutes a year. TÜV Rheinland has tested, audited and certified the process that lies behind this initiative. The certification of a process is based on the assumption that a well-organized process that is understood, supported and put into practice by everyone concerned will always produce the same result. This aim of the Zero Outage program is to limit cloud downtime for large T-Systems customers to 0.001 percent, or preferably, zero percent.

Just ten years ago, the most common cause of system failure was loss of electrical power. In Germany, power outages on the medium-voltage grid amount to around ten hours per year, the best value in Europe. Blackouts covering entire regions, such as those experienced in California, Virginia and other states in the US every year as a result of natural disasters, are unknown. However, due to globalization, they now affect us indirectly.

Just imagine a situation in which a major European telecoms company bills customers for calls to the exact second and uses the data center in Ashburn, Virginia, run by Amazon (the same data center that was crippled by a huge storm in 2012). If the telecoms company is unable to calculate the duration of customer calls, the drop in revenues could easily reach seven figures in just an hour, since only its billing capability is affected and not the telephone service itself.

During the past 20 years, there has been no slowdown in the speed of technological development or in the demand for storage. So why should the next 20 years be any different? In the final analysis, the customer does not care whether the loss of service was caused by a virus, a cyber attack, a staff member working on the inside, or a power outage.

11.4   The Role of the IT Provider

What does this trend mean for companies using cloud services and for those who offer these services? Does more effort have to be made to double and redouble data security in complex systems? Or should we return to embedded systems that act autonomously without any reference to the "outside world"? Or is there a third way – and if so, what does it look like?

The following observation is very interesting from a psychological point of view: If a company stores all of its data on its own computers in its own data center with its own IT staff, it will tolerate some downtime. There are hardly any companies able to guarantee 99.999 percent availability of their own IT systems for their employees. 99 or even 99.9 percent would be considered a success. Yet if the same company decides to outsource some of its data to cloud service providers, they often expect 100 percent availability. The reasoning is: "Why would I outsource my data if it wasn't cheaper, better and more reliable?"

There is another side to this. People have a tendency to be trusting, and when things are complex and difficult for the individual to understand, to have "blind" faith in others. This applies whether you are spending money on food, car accessories, textiles, the safety of a power plant or the confidentiality of a banking advisor. We want to trust, and that is a good thing. This can lead to great disappointment if it ultimately transpires that your trust was misplaced.

No one can deny today that the cloud is a growth driver. The fact that we need more rather than less cloud is a consequence of this. But Wikileaks, Edward Snowden and the NSA scandal have shaken our confidence in the security of data.

11.5 More Data, More Cloud Means Greater Demands on Security

This is the trade-off we have to deal with. At the same time – at least in the view of those in charge of Google – B2B and B2C models have now merged into a homogeneous B2P (business-to-people) model. The divisions between private and public, business and personal are being broken down.

In the era of the smartphone, the customer is king and the device is the scepter with which he can directly and often quite emotionally inform the entire world of his frustration with the "failure" of a manufacturer or service provider. And the tone in the business community appears to be harsher and, above all, more personal.

You don't have to think too hard, however, to realize that you cannot delegate responsibility for data – especially if it is an integral part of your business – to third parties entirely without risk. After all, there is no such thing as 100 percent security, and the costs of providing security rise steeply the closer you get to 100 percent.

For example, are the apples we buy in the supermarket safe? How can we be sure that nobody has injected them with something harmful or sprayed them with poison? To guarantee 100 percent safety, we would have to cut each apple open before selling it, examine it, and then unfortunately dispose of it because it would no longer be fit for sale. So we have to balance our confidence in the supermarket with the possible risks of buying poisoned food.

When the unexpected happens, there is an unfortunate tendency in our culture to immediately look for someone to blame. We are often more interested in assigning guilt than trying to learn how to prevent such things from happening again in the future.

So, we should not be concerned if the growth in the size of IT infrastructures is making businesses more vulnerable, because the risks can be minimized through the use of the appropriate security technology. We need to be aware of one thing, however: tomorrow's successful business models could not have happened without the cloud, as the next four examples show.

- Uber is the world's most successful taxi company. It doesn't own a single taxi, but it does have a lot of IT infrastructure.

- Facebook is the world's largest media company. It has no content of its own, but hosts content provided by its users.

- The Chinese company Alibaba is the world's largest trading company, although it does not manufacture any products or have any warehouse space.

- The world's largest accommodation company is Airbnb, which does not own any property.

What all of these four companies have in common is that they rely on your confidence in their brand and your confidence that your data is safe with them. And that is not always easy to achieve. After all, the countries in which these companies base their data centers all have completely different legal standards for the protection of data. Despite this, the companies have to guarantee 24/7 availability on all types of devices and applications anywhere in the world. That is what you call B2P! Here, the expectations of the users of the service align with the expectations of the companies in terms of IT system stability, or "Zero Outage," and cloud security.

What cannot be delegated to the cloud, or to providers such as T-Systems, is the responsibility for one's own commercial success. B2P users of the future will be as powerless to delegate responsibility for supplying their own data as Alibaba or Facebook. As long as we are clear about that, we will also accept 99.999 percent uptime and realize that 0.001 percent downtime is perfectly reasonable.

11.6   Conclusion

The cloud is like the invention of printing. All of a sudden, people have access to knowledge about the world, but to make use of it, they have to be able to read (and understand). Programs like "Zero Outage" from T-Systems do not – to continue with this simile – teach you how to read, but rather give you confidence in the reading matter. Then it is up to you (or your company) to decide what to do with the information. Being human means making mistakes – but learning from them in order to improve.

Michael Weppler is Executive Vice President Systems at TÜV Rheinland, one of the world's leading providers of testing services. He is responsible for its worldwide management certifications business and revenue of more than 150 million euros. Michael Weppler is 50 years old and was born in Frankenthal. After an apprenticeship as a biology lab technician at BASF, he attended the University of Applied Sciences of Rhineland-Palatinate, where he studied process engineering, specializing in biotechnology. In the late 1990s, he joined TÜV Rheinland's Mobility business unit, where he held various management positions before being promoted to Executive Vice President Systems in 2012. The Systems division appraises management systems, IT processes and entire companies in accordance with internationally recognized standards and specific performance criteria. The division's specialists act as independent third parties to certify that companies are systematically and consistently implementing and complying with predefined standards.

Conclusion and Outlook

12

Ferri Abolhassan

What is driving digitalization today and why is the cloud such an integral part of it? The analyses in this book by respected and experienced experts from industry, business and the media show that the disruptive force of digitalization is radically changing the way in which business works. Many analog business models and processes are being replaced by digital concepts. The transformation has been affecting all business sectors for some time – from publishing and the music industry, to commerce, manufacturing and logistics (cf. Kempf 2014).

Thanks to streaming, the music industry now operates under a completely new set of rules. The success of Uber is a good example of how an entire industry – in this case, the taxi industry – is currently experiencing its own digital tsunami. Taxi companies are trying to defend their traditional turf by formal, legal means. But anyone who wants to play the digitalization game today needs a forward-looking strategy.

Regardless of whether we are talking about apps used simultaneously by millions of people, or improving medical care with intelligent pills, or mobile applications that deliver safety-critical realtime information, or breaking down local IT barriers so that staff can work together efficiently across national borders: All of this relies on the cloud to function. Only the cloud can collect and centrally store the infinite amounts of data being generated by the Internet of Things, and then make use of that data with the aid of big data technologies. The cloud's capabilities are making it a key factor in tomorrow's digitalized businesses, processes and products. It has the requisite speed, intelligence and flexibility, and it has the capacity and scalability. Managing such growing complexity and radical change requires people with many years of highly advanced technological and consulting expertise. Specialists with these skills are therefore in demand. After all, it takes a great deal of experience to take an entire legacy infrastructure to the cloud. As tasks

F. Abolhassan (✉)
T-Systems International GmbH, Mecklenburgring 25, 66121 Saarbrücken, Germany
e-mail: Ferri.Abolhassan@t-systems.com

© Springer International Publishing Switzerland 2017
F. Abolhassan (ed.), *The Drivers of Digital Transformation*, Management for Professionals, DOI 10.1007/978-3-319-31824-0_12

go, this one has more in common with open heart surgery than taking a walk in the park.

This book deals with this specific challenge and also explores the opportunities of digital transformation. It provides an exhaustive 360-degree view of the entire topic – from the organizational and technological fundamentals, to requirements in areas such as security, quality and standardization, through to real-life practical examples. It provides practical guidance for CIOs and IT managers, as well as for CEOs, on how to optimize their digital business models and carry out the transformation their businesses urgently need.

This book charts a path and provides clear directions. There are three central propositions that point to the direction that future trends are likely to take:

12.1   The Cloud Is Normality and the Market Is Growing

The examples provided in the various chapters show that the cloud has become an integral part of normality for consumers and businesses alike. It is impossible to imagine many business or consumer applications – not to mention business processes – functioning without the cloud. All business sectors can benefit from the power of the cloud and the flexibility and scalability it offers. The examples given in this book demonstrate this unequivocally. This trend will continue. The cloud will grow in importance for both businesses and consumers in the future.

While today's technologies and applications may have been inconceivable ten years ago, the innovative power of the cloud has now changed whole industries and created new ones. We can only speculate about where this trend is heading, how fast it will develop, and which innovations will shape society in the future. What we do know for certain is that the importance of the cloud and digitalization continues to increase across all industries. In the ICT industry alone, it has made tremendous strides. Seven out of ten (71 percent) ICT companies currently (as at 2015) use cloud solutions (cf. Bitkom 2015). This is a seven percent increase over the previous year. The more traditional sectors such as the auto industry, banks, insurance companies or chemicals and pharmaceuticals are also expanding their use of cloud solutions. This trend will carry on over the next few years and contribute to growth in the size of the market and in the number of potential uses for cloud solutions.

12.2   The Cloud Is and Will Remain a Collaborative World

The developments over recent years have revealed a fundamental truth: Nothing happens without partnerships. The market is not only experiencing a steady upward trend, it is also extremely dynamic. But enterprises need the relevant expertise and the right partner if they want their customers to benefit from improved quality, flexibility and efficiency in their fast-moving daily business and if they want to successfully tackle the challenges they will face tomorrow. Digitalization and the transformation to the cloud require technology and expertise together with the manpower capable of delivering highly complex solutions and services. It is not a simple matter to transfer, for example, tens of thousands of employees or thousands of applications to the cloud. It requires specialist support. It needs experts with years of experience in the digitalization and optimization of business processes – experts who are also familiar with the technical requirements and implementation options. These partners rely in turn on an entire partner ecosystem that includes hardware and software suppliers. Technologies are changing so rapidly today that it takes a value-creating collaborative network to transform businesses safely and reliably.

12.3 The Cloud Must Be Simple, Reliable and Affordable

Regardless of where companies are on their own road to digital transformation, there are some key questions that they should always ask. What requirements does the cloud have to meet within the company? Where will it create business value, competitive advantage or improved customer loyalty? How can cloud technology improve efficiency and reduce costs? Then there is the question of finding the right form of cloud. Do you need a public, private or a hybrid cloud? Here, CIOs are increasingly differentiating between commodity and non-commodity IT. Other questions to ask are: Which business processes contain data-sensitive workflows or information and require special attention in terms of security and quality? Or, how can we provide, for example, a stylish, lean app that delivers a straightforward and positive user experience while at the same time ensuring the security of customer data?

Whatever the conclusion each individual company reaches on these matters, it is evident that, in addition to maximum scalability and performance, the cloud must meet three criteria both now and in the future: It must be simple, it must be secure and it must be affordable – for large enterprises as well as for small and mid-sized companies and for end customers.

Equally important, the technology must evolve in line with user expectations. Solutions taking the first step towards delivering integrated solutions to clients include software-defined data centers, which offer users IT-as-a-service on demand – along with, of course, high reliability, availability and data security. And efficiency – which is why standardization and automation continue to increase.

A look at the current state of play and trends for the future make it clear that cloud is a major issue that will ultimately impact businesses in all industries. In spite of this, some 25 percent of German IT decision-makers still do not have the cloud on their agenda (cf. Büst and Crisp Research 2015). This book provides some useful starting points and principles for getting to grips with the tasks that lie ahead and moving successfully into the future. The first step has been taken. It is now up to all those with the capacity to do so to accept the responsibility of taking the next steps that will lead towards a profitable future.

References

Bitkom (2015). *"Cloud-Monitor 2015" by Bitkom Research and Bitkom on behalf of KPMG*. Accessed August 19, 2015, from https://www.bitkom.org/Presse/Presseinformation/7-von-10-IT-Unternehmen-setzen-auf-Cloud-Technologien.html

Büst, T., & Crisp Research (2015). *Cloud-Market Update 2015: Wolkig mit Aussichten auf digitale Unternehmen*. Accessed November 09, 2015, from http://www.crisp-research.com/cloud-market-update-2015-wolkig-mit-aussichten-auf-digitale-unternehmen/

Kempf, D., & Bitkom (2014). Accessed August 09, 2015, from https://www.bitkom.org/Bitkom/Blog/Blog-Seiten_1780.html

Ferri Abolhassan Ferri Abolhassan, a computer science graduate, secured his first professional role as part of Siemens' R&D team in Munich. He then worked at IBM in San Jose, USA. In 1992 he joined software vendor SAP, remaining until 2001. Abolhassan held a number of senior positions during this period, including a spell as Senior Vice President of the global Retail Solutions business unit. Following a 4-year tenure as Co-CEO and Co-Chairman at IDS Scheer, he returned to SAP in 2005, most recently as Executive Vice President, Large Enterprise for EMEA.

In 2008, Abolhassan moved to T-Systems, where he became Head of the new unit Systems Integration and joined the T-Systems Board of Management. In late 2010, Abolhassan took on role of Head of Production, before becoming Director of Delivery in 2013. Abolhassan was appointed Director of the IT Division in 2015, overseeing approximately 30,000 employees and 6,000 customers. In addition to his current function, Abolhassan has been responsible since late 2015 for building up the new business division "Telekom Security". The new unit will consolidate the security departments from all different Group units of Deutsche Telekom.

Printed in the United States
By Bookmasters